HARVEST OF HAZARDS

IOWA AND THE MIDWEST EXPERIENCE
Series editor, William B. Friedricks,
Iowa History Center at Simpson College

DEREK S. ODEN

HARVEST OF HAZARDS

FAMILY FARMING, ACCIDENTS, AND EXPERTISE IN THE CORN BELT, 1940–1975

UNIVERSITY OF IOWA PRESS, IOWA CITY

UNIVERSITY OF IOWA PRESS, IOWA CITY 52242

www.uipress.uiowa.edu
Printed in the United States of America

DESIGN BY TERESA W. WINGFIELD

The University of Iowa Press is a member of Green Press Initiative and is committed to preserving natural resources.

Printed on acid-free paper

LIBRARY OF CONGRESS CATALOGING-IN-PUBLICATION DATA
Names: Oden, Derek, 1972– author.
Title: Harvest of hazards : family farming, accidents, and expertise in the Corn Belt, 1940–1975 / Derek Oden.
Other titles: Iowa and the Midwest experience.
Description: Iowa City : University of Iowa Press, [2017] | Series: Iowa and the Midwest experience | Includes index.
Identifiers: LCCN 2016035921 | ISBN 978-1-60938-498-2 (pbk) | ISBN 978-1-60938-499-9 (ebk)
Subjects: LCSH: Agriculture—Middle West—Safety measures. | Agriculture—Middle West—History—20th century. | Family farms—Middle West—Safety measures. | Family farms—Middle West—History—20th century. | Industrial accidents—Middle West—History—20th century.
Classification: LCC S565 .O34 2017 | DDC 630.28/90977—dc23
LC record available at https://lccn.loc.gov/2016035921

For my children

Luke, Caleb, and Cora

and for my wife, Jennifer

CONTENTS

ACKNOWLEDGMENTS

THE NUMBER OF HISTORY FACULTY to whom I am indebted are many; but several deserve special recognition. Dr. Harl Dalstrom at the University of Nebraska–Omaha encouraged my initial interest in the study of the Midwest. I first became acquainted with the issue of safety when writing my master's thesis. I was studying the coal mines of Southern Iowa at the time, and little did I know this research would lead to investigating the physical risks of farming. I would also like to express particular gratitude to the late Dr. Hamilton Cravens, whom I came to know and greatly admire. His professional guidance and mentoring were crucial in my development while at Iowa State University. His contributions to the initial stages of this work were immeasurable, as were his continued encouragement, friendship, and kindness. I also benefited from Dr. Pamela Riney-Kehrberg's scholarly guidance. Her comments were vital in helping me grapple with the complicated topic of farm accidents.

I was very blessed to gain the support of many people and institutions over the many years that accumulated while finishing the project. I benefited greatly from the friendship and advice of many fellow graduate students during the early stages of this project. I am also very grateful to Iowa State University's Special Collections and University Archives staff. These rich collections were indispensable in the book's completion. I am especially thankful to my colleague Dr. Mark Robbins of Del Mar College who encouraged me to resuscitate my dissertation and transform it into a book. I am also greatly indebted to those who helped me locate midwesterners who were enthusiastic about sharing their agricultural experiences. Mary Ellen Schmitz, Steven Johnson, and Marilyn Walker were all very helpful in putting me in contact with knowledgeable and generous people. It was a privilege to be welcomed into the homes of farm families who provided indispensable insights and generous portions of hot coffee over several wintry days. Their warmth and kindness constitute some of my favorite memories of the project.

The State Historical Society of Iowa (SHSI) merits special recognition in supporting my work. I was able to travel to Iowa from my adopted state of Texas to conduct interviews because of a generous SHSI research grant. Additionally, the SHSI allowed me to republish portions of an article I composed for the *Annals of Iowa*. They were also very helpful in quickly granting me copyright permission for the photographs that greatly enhance the work. The Agricultural History Society also deserves my appreciation for granting me permission to incorporate parts of an article that appeared in *Agricultural History*. The feedback I received from anonymous reviewers while completing both of these pieces provided important critiques at crucial junctures.

I would also like to thank the staff of the University of Iowa Press, Simpson College's Iowa History Center, as well as all those who support the Midwest Experience book series. The publication of this book would have been impossible without them. I particularly wish to thank Dr. Catherine Cocks, who believed in this project from beginning to end. Her judicious critiques, patience, and encouragement were consistent throughout the years. Dr. William Friedricks's belief in the project's value was heartening even when I began to grow weary of the process. His suggestions greatly enhanced the book's final form. The anonymous readers also played a critical role in improving this work, and I am grateful for their suggestions. The suggestion to reorganize the book's structure was particularly potent and greatly enhanced the work's readability. Any remaining mistakes are purely my own.

Finally, I wish to thank my parents, Malcolm and Susie Oden, whose love and support have been constant through my life. My father talked to me about baling hay long before I understood what that really meant, while my grandma Lucy told me stories of mining even before I entered school. Some of these recollections included accidents involving severe injury and even death, but all were seasoned with calls to perseverance and courage. I greatly admire the traditions of devotion to work and family that have been displayed by Corn Belt families for generations. My wife and children have in their own way adopted these estimable customs. They patiently endured my absence as I simultaneously taught summer classes and completed this book. My wife Jennifer's sacrifices and contributions were particularly great. She not only provided abundant emotional support but her wordsmith talents and computer skills were welcome antidotes in moments of frustration and fatigue. My wife deserves the greatest thanks for being a constant, reassuring presence from the work's germination to its successful end.

HARVEST OF HAZARDS

1

FROM ACCIDENTS TO ACTION

THE HAZARDS PRESENT on Corn Belt farms during the twentieth century have long been recognized. Jim Klein, who grew up on a farm near Ashton in northern Iowa in the 1960s and 1970s, commented on farming's many dangers. After reflecting on his youth, he stated, "There were at least a dozen times where I should have been killed."[1] Larry Glenn, a lifelong Iowa farmer in Wapello County, Iowa, expressed similar sentiments, noting one had to be careful around "everything."[2] This should not be surprising, since Corn Belt farmers work under conditions distinct from those of workers in other industries, including farmers' entrepreneurial status, their rural locations, and the variety of technologies required to run their farms. Small operators have also dominated the region's agriculture, making intensive safety legislation, company measures, and state inspection systems impractical. Where workers in other industrial sectors often possess unions to advocate on their behalf, achieving various protections, farmers, being self-employed, did not benefit from them, or from the workers' compensation and similar laws enacted in most states between 1910 and 1921.[3] Such realities meant that throughout the entire mid-twentieth century farm safety educational programs represented a crucial component of the effort to improve agricultural safety.

Corn Belt farm families resided farther away from emergency care providers, a particularly distinctive aspect of their work and one that amplified the dangers of their agricultural lifestyle. Ruth Falck, who lived on a northern Iowa dairy farm, was fully cognizant of her family's vulnerability related to its distance from medical help, and she acquired first-aid training so she would be better prepared to meet the farm's many hazards.[4] James Wilson from Appanoose County, Iowa, highlighted this aspect as well in describing his accident involving a hay baler. He stressed that it took a long time for his dad to get him to a medical facility, as it was a considerable distance to the hospital.[5] Sadly, such isolation meant that farmers often worked and sometimes even died alone. Laverne Jacobson was killed while discing with his tractor late in the evening on May 7, 1969, near Nevada, Iowa. Investigators believed he fell from the tractor and was run over and dragged by the wheels and attached disc before the tractor stalled. His wife was horrified when she found him dead later that night, "caught between the tractor and the disc."[6] Such stories highlight the ever-present risks associated with farmwork; unfortunately, few historians have investigated these dangers or the vigorous safety movement that has worked to reduce these hazards.[7]

Historians have done much to examine other aspects of the impact of technological innovation in Corn Belt farming, however. For instance, J. L. Anderson made insightful observations regarding farmers' adoption of technology in his work *Industrializing the Corn Belt: Agriculture, Technology, and Environment, 1945–1972*. He emphasized that farmers created a new reality by rapidly acquiring gasoline-powered machines and a new repertoire of chemicals and grain-handling equipment. Anderson stressed that by the late twentieth century, "[t]hese physical manifestations of technological change were signs that Iowa, the heart of the Corn Belt, was an industrial landscape as much as it was a rural one."[8] Although such conclusions applied to the entire Corn Belt, as well as to many other agricultural regions, Anderson adopted a state-level approach, focusing on Iowa as the "leading state in feed grain and livestock production," one of the most cultivated states in the nation, and one dominated by "family farming" throughout the period.[9]

Like Anderson, I discuss national trends, but with a focus on the agricultural classification of the Midwest known as the Corn Belt.[10] Agricultural historian R. Douglas Hurt defines the Corn Belt as "the portion of the Midwest where farmers emphasized that crop and produced bountiful harvests." He goes on to state that by the 1980s, it "extended from central Ohio on the east to Nebraska on the West."[11] Geographer John C. Hudson, author of *Making the Corn Belt: A Geographical History of Middle-Western Agriculture*, places the origin of the idea of a specific area known as the Corn Belt even before the current conception of the Midwest had materialized, pointing out that the area has been recognized as a subregion of the United States for over a century.[12] The Corn Belt is an ideal place to investigate the farm safety story, since safety activists from the area played a leading role in the national effort to make farms safer. Midwestern farm safety experts, state-level farm safety councils, and 4-H leaders consistently made important contributions throughout the last half of the twentieth century to efforts to address the issue.

Technology figures centrally in the story of twentieth-century farm hazards, as it does in the story of twentieth-century farms. Anderson argued that farmers were the primary decision makers as to how and to what degree they specialized to meet the demands of the marketplace, both in streamlining their methods and in adopting new technologies. Such approaches resulted in a national trend of operators dependent on both monoculture and capital-intensive technologies to boost production.[13] Americans living through and participating in these changes were fully aware of this great transformation. In 1941, a rural journalist in Illinois wrote, "Some farmers believe that this is fast developing into a machine age in farming and that within a few years horses will be little used on farms and some places not at all."[14] Corn Belt farms were becoming as much an outdoor factory as an organic process consisting of sun, soil, muscle power, and manure.

This process of agricultural industrialization introduced new dangers for those living and working on the nation's farmsteads, just as it had done in many other sectors of the economy. Innovations such as new kinds of machinery, storage equipment, and chemicals created a host of safety challenges for rural residents. Even electricity, adopted by all Americans, appears to have held special risks for the farming population. For instance, those living in the countryside often used electricity both in their households and in their farming operations. Their use of the power source for heat and fencing, the increasing presence of electric lines, and ever-larger agricultural equipment all posed exceptional risks to farmers. Improperly installed heat lamps could easily ignite dry hay, a carelessly installed electric fence exposed individuals to the possibility of electrocution, and farm equipment operators might accidentally contact overhead electrical cables. Such hazards were largely unknown to urbanites.

In the early days of the 1940s, the movement centered more on education than on legislation or engineering improvements. A diverse group of governmental, business, and nonprofit organizations worked together to create an educational effort targeting an impressive variety of rural safety issues, and each participant fulfilled an important role in the collective effort to decrease accidents. For example, the National Safety Council (NSC) and the United States Department of Agriculture (USDA) provided administrative leadership on a national level, while 4-H chapters and other entities delivered local educational programming. A varied group including large corporations, small businesses, and other private interests provided the financial support required for farm safety activities to address a variety of farm hazards.

Safety concerns related to a range of issues, including use of agricultural machinery, chemicals, and a variety of other factors contributing to farm injuries and fatalities. Success in addressing such a variety of concerns required collective efforts by a diverse group of organizations. Cooperative programs, such as National Farm Safety Week, symbolize the movement's harmonious tone. Rapid development of state farm safety committees throughout the nation also demonstrates the accommodating, inclusive atmosphere in the early years. The rural population's enthusiastic participation in farm safety programs and the private sector's enthusiastic financial support of such efforts provide additional examples of this relatively unified environment.

However, this cooperative atmosphere had changed by the late 1960s and early 1970s. Federal regulatory interventions such as the Occupational Safety and Health Administration's (OSHA) relatively timid but nonetheless largely unpopular actions showcased how large segments of the agricultural community were antagonistic toward governmental intervention. The appearance of the farm safety professional which had begun at the outset of the movement became fully realized by the early 1960s. The creation of a separate National Institute of Farm Safety (NIFS) organization symbolizes how this highly trained group of safety professionals desired to improve their occupation's expert reputation. Such efforts appear to have contributed to a general fragmentation of a movement which had been relatively unified during the 1940s. The growing number of farm accident liability cases reveals how Americans increasingly pointed to manufacturers of all kinds of products in a variety of industries as largely to blame for their accidents.[15] This trend created divisions among agricultural engineers, farm safety experts, equipment manufacturers, and other groups who had been involved in the safety effort. Thus, this study ends in the 1970s, when dramatic changes had significantly altered the politics and tenor of the farm safety movement. (A treatment of the topic in the last three decades of the twentieth century could best be handled in a separate work focusing on that period's unique aspects.) The aim here is to provide greater historical context for mid-twentieth-century farm safety efforts. To understand the issue, it is first critical to understand the occupational distinctiveness of farming in the Corn Belt.

During the period of this study, farmers were entrepreneurs, which distinguished them from workers in the manufacturing or extractive industries, such as coal mines. Some farmers viewed this aspect of their chosen career as beneficial, offering greater flexibility and freedom than that found in other professions. John Van Ringelestein, of Wapello County, Iowa, said that being his "own boss" was an advantage of the agricultural lifestyle.[16] Prospective farmers during the period made similar comments regarding the independence of the farming way of life. In 1950, Dwight Colliver, a seventeen-year-old FFA member from Sigourney, Iowa, stressed that his interest in farming was largely related to the great degree of autonomy it offered in comparison to other occupations.[17] In 1966, Larry A. Petersen, a farmer in Plymouth County, near Merrill, Iowa, highlighted the fact that "working for [him]self" was preferable to having a boss.[18] Despite the advantages of such autonomy, farmers faced considerable challenges during the era that proved to be daunting even to the most skilled agriculturalists.

Although the Second World War created massive demand for agricultural commodities, farmers also faced increasing production costs throughout that period and into the following decades. For example, a farmer buying a John Deere Model 620 spent $2,100 in 1958. A mere seven years later, he would pay $4,169 for "J. L. Case's top-of-line 401 diesel general-purpose tractor." By 1977, Case's least expensive tractor cost $15,000.[19] Farmers also chose from a complex menu of technological choices during the era. For instance, by 1972, historians Dennis S. Nordin and Roy V. Scott indicated that farmers could select from "4,220 different formulas to kill worms and insects." Such figures dramatically indicate how much the profession was changing technologically.[20] It was an era of considerable financial stress during which many opted to abandon agriculture. Nordin and Scott state that farmers who made the correct financial decisions with regard to land and machinery persisted, but "for many more . . . careers on farms would end involuntarily between 1945 and 1970."[21] Keeping themselves and their families safe was only one of many concerns farmers encountered during a period of dramatic change.

Such complexity created a multitude of new challenges, and a variety of experts sought to market and apply their solutions to meet them.

The growing influence of experts manifested itself in a rising chorus calling for additional professionals to refit rural life to the modern age. This cultural impulse had long since been reflected in the increasing influence experts exerted in many areas of American society and culture. Scholars such as Burton J. Bledstein, Gerald N. Grob, Allan V. Horwitz, and many others have explored the pervasive influence of expertise from the late nineteenth century to the present.[22] Burton Bledstein's comments are especially useful in explaining the origins and features of this late nineteenth-century development. He asserts, "In a nation without an effective apprenticeship system and without a significant gentry, the school diploma more and more served as the license with which an individual sought entry into the respectability and rewards of a profession." Thus, according to Bledstein, professional attainment and economic advancement became ever more contingent on one's ability to provide a specific form of expertise considered valuable and useful to the wider public.[23] Such trends contributed to the development of a stronger sense of professional identity across diverse fields, including education, dentistry, and library science.[24] Bledstein also argues that these professionals expended considerable energies in establishing professional legitimacy by working "to define a coherent system of necessary knowledge within a precise territory, and to control the intrinsic relationships of their subject by making it a scholarly as well as applied science."[25] The culture of expertise would influence many aspects of American life—including farm safety—by the Second World War.

Unlike the industrial managerial class who had exerted their authority over factory workers, however, experts such as agricultural engineers, soil scientists, and entomologists could in most cases only educate, not mandate, modernization.[26] Nonetheless, the growing power of agricultural experts represented an increasingly bureaucratic and urban orientation of national life. Historian David Danbom emphasized that one result of this change was "farmers had become peculiar. They were objects of concern. They drew the attention of altruistic reformers who suggested urban innovations for their institutions and solutions for their problems."[27] Such observations show that although farmers valued their independence, it may have been harder to maintain as their jobs became increasingly intricate and technologically sophisticated. One might surmise that farmers' beliefs in their autonomy may have become more illusionary than real by the mid-twentieth century as financial pressures, government involvement, and persistent technological change exerted tremendous influence on family farms in the Corn Belt.

Although this study begins with the onset of the Second World War, this in no way is meant to suggest that farming was without risk prior to the mid-twentieth century or that no efforts to improve farm safety had existed prior to the war. In fact, farming has always possessed some level of danger, and death has most likely always been a grim possibility for farmers. Farm families in the Corn Belt have always faced potential harm from unpredictable animals; sharp, fast-moving tools; and nature's vagaries. Such realities had been present long before the widespread use of gasoline-powered equipment, agricultural chemicals, and large grain storage facilities. The Second World War marks an important transitional period in farm safety, however, because farm families were no longer largely alone in dealing with the problem. The war brought a national mobilization effort and encouraged development of a more organized approach to farm safety through formal, bureaucratic, and technologically sophisticated solutions.[28]

Sadly, farmers could be killed or injured while using the most basic tools, such as pitchforks. Sharp hand tools could inflict ghastly injuries or fatalities, as revealed in many stories circulating throughout the Midwest. In 1943, Claire Swisher, editor of the *Hammermill Bond,* wrote about the dangers of pitchforks. In 1945, a *Farm Safety Review* writer described the fate of Mr. Hedge, a farmer from Leavenworth, Kansas, who "climbed onto the hay loft, threw food down for the cattle, and tossed the fork into it. As he started down the ladder, he missed the top rung. The additional impact broke the second rung and he fell. The fork handle pierced deeply into his abdomen, and he died a few days later."[29] Larry Glenn remembered an uncle telling him about a pitchfork accident involving a youth who was playing while others were putting up loose hay: The boy was impaled.[30] Glenn said that such stories left a powerful impression on him of the hazards of farmwork, even from some of the profession's simplest and oldest tools.

Not only had farmers experienced hazards in their work long before the Second World War, but some individuals had already raised awareness about the problem. By the time of the First World War, literature on the subject and on the need for training to adapt to technological change was beginning to emerge, albeit in a slow and isolated fashion. In 1918, an article in the *Postville Herald* entitled "Tractor Gaining in Popularity in the West" recognized that farmers' decisions to procure tractors in an ongoing process of mechanization created the need for operational training. The author encouraged farmers to "take a course of training under competent instructors" to refine their skill with the new machine.[31] Such comments foreshadowed the view of many mid-twentieth-century safety advocates that farmers needed assistance from outsiders to adapt to the many technological changes that were rapidly affecting agricultural work.

Agricultural equipment manufacturers and state-level youth groups had also already demonstrated an interest in improving farm safety before the Second World War. In 1934, T. A. Erickson, a Minnesota Extension official, believed that farm accidents represented a very real problem; thus, he successfully encouraged 4-H clubs to integrate safety into their programming. These activities included "safety contests" and other efforts to promote safer farm practices.[32] In 1938, the members of the Farm Equipment Institute, an association of the nation's agricultural equipment manufacturers, established a safety committee to advise its associates regarding the problem. Frank H. Harrison, a high-ranking International Harvester Company official and safety committee member, summarized the group's initial activities. He indicated that progress had been made in developing standards for safety devices, improving equipment instruction handbooks, and promoting operator safety in other ways. Despite such pioneering efforts, concerns regarding the need for sufficient agricultural laborers, coupled with the desire to use these resources efficiently, greatly intensified safety efforts.[33]

The National Safety Council (NSC) was particularly important in providing leadership in the mounting effort to reduce the number of farm accidents. In 1938, the NSC leadership called for a meeting of agricultural leaders to initiate efforts to improve safety on the farm. C. M. Segreaves, a safety director for the Illinois Agricultural Association, underlined the importance of the division's first gathering, believing that the issue had been ignored for much too long. Segreaves stated that "although material on almost every conceivable phase of rural life has been carefully compiled by the United States Department of Agriculture, taught in our agricultural colleges and disseminated to farmers through the extension service, we find little or nothing on safety. Control of grasshoppers, clinch bugs, Bangs disease, and noxious weeds . . . yes! Control and prevention of farm accidents . . . no!"[34] Segreaves and other initial meeting attendees hoped to bring attention to the long-neglected problem of farm safety.

In the 1970s, Iowa State University agricultural engineer Dale Hull also stressed that safety efforts had materialized prior to the war, noting that Iowa 4-H clubs were receiving materials from the extension service regarding fire and electrical hazards before that time. Nevertheless, Hull emphasized, "Entry of the United States into World War II had a very definite effect on the agricultural engineering extension programs," especially in their emphasis on safety.[35] The author of a study from the USDA Bureau of Agricultural Economics of the farm accident problem concurred, stating, "With few exceptions, little organized work in either accident prevention or accident insurance has been undertaken for farmworkers" as of the year 1942.[36] Such observations acknowledge that although some work had been devoted to the issue prior to the war, efforts greatly intensified in the 1940s.

Although farmers were not alone in laboring in a potentially dangerous environment, they were slow to experience accident prevention improvements. Workers in most other industries experienced reductions in work-related hazards long before farmers. Governmental regulation, the pressure of organized labor, and voluntary corporate efforts combined to drastically improve workplace safety. As productivity in industrial occupations increased, large numbers of individuals left these comparatively dangerous occupations, contributing toward the funneling of workers into safer occupations. Agriculture provides an example. The rural exodus accelerated during the era, as new technologies meant fewer people were needed on farms. This held for the various extractive industries and for other relatively dangerous professions as well. The difference lies in the level of danger faced by those left behind: Farm families who remained experienced little improvement in their working conditions,[37] while for the reasons sketched above, workplace safety elsewhere improved.

Railroad laborers were some of the first employees to witness workplace safety improvements. In the late nineteenth century, railroad workers were exposed to a variety of hazards, including boiler explosions, crashes, and rail-car coupling accidents. The collective efforts of the federal and state governments as well as the voluntary efforts of railroad companies improved the safety record. In fact, railroad workers received federal protection decades before legislative safeguards were established for farmworkers. In 1893, Congress enacted the Federal Safety Appliance Act requiring railroads to install features such as "air brakes, automatic couplers, and handholds on freight cars."[38]

Unfortunately, the severity and frequency of railroad accidents persisted despite new regulations and technical improvements. Interstate Commerce Commission (ICC) officials responded by researching the issue, proposing innovative solutions, and supporting governmental regulation. The Hours of Service Act represented one piece of legislation aimed at reducing worker fatigue and consequently decreasing accidents. In 1908, Congress also enacted of the Ash Pan Law to improve safety by reducing railroad workers' exposure to hot steam engine ash. ICC representatives devoted additional resources to railroad inspections to check for safety law compliance. Industry leaders also improved rail quality and communication technologies, both of which reduced derailments and collisions.[39]

The men who labored to feed the nation's insatiable appetite for coal were also injured and killed in a myriad of ways. Some of the more prominent incidents involved death from explosions, roof cave-ins, or coal-car accidents. A particularly well known mine catastrophe occurred at Monongah, West Virginia, and left 361 miners dead, testifying to mining's extreme dangers. Safety advocates were unable to reduce mine fatalities to the degree they had reduced fatalities in the railroad and manufacturing sectors. Nevertheless, a combination of governmental intervention, union activities, and employer contributions eventually improved working conditions. State governments were some of the first to enact laws requiring basic safety provisions and the hiring of individuals to make compliance investigations. Unfortunately, state regulations were often not sufficiently comprehensive in covering the spectrum of potential mining hazards, including the widespread incidence of roof collapse. Issues such as confused wording in the laws and uneven enforcement slowed progress further. The effectiveness of state law was also reduced because employers' fines for lack of compliance were often small.[40]

Although the movement to improve mine safety was paltry in comparison to similar movements in other industrial sectors, miners benefited from a much greater degree of safety intervention than did farmers. Miners were the benefactors of federal and state institutions, as well as of efforts by the United Mine Workers (UMW) union. The Bureau of Mine officials researched mine accidents, established safety equipment standards, and disseminated safety information. UMW representatives often alerted state mine inspectors to hazards at particularly dangerous mines and recommended that their members use safety equipment. Mine owners also became aware of the advantages of reducing accidents and instituted a variety of safety measures. Company efforts included restructuring mine layouts with a greater degree of safety awareness, publishing safety manuals, and reprimanding miners who worked unsafely.[41] Farm families worked throughout the early twentieth century without the benefit of such corporate safety activities.

Factory laborers also benefited from a mixture of safety legislation, union activity, and voluntary company efforts. Although the nation's manufacturing companies greatly expanded their productive capabilities, industrialization also came at a cost. The factory environment contained a myriad of potential hazards, including pressurized boilers, fast-moving belts, and presses, all of which could potentially kill or maim. Workers nevertheless experienced dramatic reductions in the dangers they faced in the workplace during the twentieth century's first decades due to collective efforts of unions, reformers, and factory owners. Samuel Gompers, legendary leader of the American Federation of Labor, vigorously supported the enactment of state worker compensation laws. State legislators also created industrial commissions to investigate work safety issues and establish codes. By 1923, over half of the nation's states possessed such commissions, which worked to reduce hazards.[42]

Manufacturers devoted a significant amount of resources to the effort as well. In 1908, leaders of U.S. steel companies established a safety committee to plan and implement safety improvement measures. The companies' leaders instigated activities such as periodic plant inspections, the prompt implementation of safety measures, and a continued re-evaluation of accident prevention measures. They soon discovered that improving safety actually enhanced the companies' profitability. Moreover, major manufacturers, such as Ford Motor Company, John Deere, and Joseph Bancroft & Sons Textile Mills, implemented similarly vigorous safety programs.[43]

A brief review of federal and state regulations regarding agri-chemicals reveals that the consumer benefited the most from agricultural chemical regulations. Such legislation again highlights how farm families continued to be vulnerable to hazards long after those in other industries had received protections. This state of affairs remained the case well into the 1970s. This was the situation since most government regulations, although far from optimal, attempted to ensure the overall quality of the farm chemicals used and to limit the amount of chemical residues on food products. Such legislation included the Federal Food and Drug Act; the Federal Cosmetic Act of 1938; the Federal Insecticide, Fungicide, and Rodenticide Act (FIFRA) of 1947; and the Food Additives Amendment of 1958. The last piece of legislation included the Delaney Amendment, which further increased the stringency of tolerance standards. Although state-level legislation eventually required formal training for both professional pesticide applicators and farmers, for most of the period, a user's willingness to follow the label's instructions was the only instrument protecting him or her from harm.[44]

The Federal Food and Drug and Cosmetic Act of 1938 represented one of the first pieces of legislation enacted to provide the American public with increased protection. The act "established procedures for setting safe limits called 'tolerances' on the amounts of pesticide residues permitted on crops." Congressional enactment in 1947 of the Federal Insecticide, Fungicide, and Rodenticide Act empowered the USDA to determine which chemicals were acceptable for use and safe for shipment across state boundaries. The act required companies that produced pesticides to "register their labels with the USDA."[45] Companies seeking to achieve the USDA's product standards were required to demonstrate that their chemical could successfully control the targeted insect, that it was safe when users followed the label's instructions, and that excessive residue levels did not remain. FDA chemists and USDA officials worked cooperatively in the registration procedures.[46]

The federal government provided the public with additional safeguards in the 1950s. In 1954, the Miller Pesticide Amendment further streamlined the process of setting tolerances, as well as decreasing potential dangers associated with eating chemically treated foods. The Miller Amendment enacted procedures essentially requiring companies to submit to rigorous self-investigation of how chemical traces would be measured and to undertake a detailed assessment of the dangerous residue levels. Following this analysis, FDA researchers analyzed the results of the companies' studies to determine safe amounts of residues. Government researchers accepted this amount of residue only "if all reasonable doubt is resolved in favor of consumer safety." If the government's scientists reached such a conclusion, the maximum allowable residue was then established for the purposes of interstate commerce.[47] One of the most important effects of the Miller Bill was that it increased the professional scrutiny involved in the chemical registration process and consequently improved food residue safeguards for consumers.[48]

In 1958, the public received additional protection with the passage of the Food Additives Amendment. This legislation also included the Delaney Amendment. In 1991, Gustave K. Kohn, an authority on pesticide regulation and a contributor to a regulatory history, stated that the Delaney Amendment "expanded protection by setting zero tolerance levels for any chemical in food in which animal meat tested at any level exhibited carcinogenicity."[49] In 1959, J. M. Magner, an entomologist employed by Monsanto, commented on the registration process. He stated that it often took five years and that the need for the chemical in fighting a particular insect must be proven. Government requirements mandated that chemical companies compose labels for their products. Federal involvement continued after the chemical's successful registration. Inspectors from the Pesticide Regulation Division, in cooperation with state administrators, bought and tested insecticides, ensuring that the chemicals' properties aligned with the label's contents and description. If it was determined that the manufacturer was negligent by misrepresenting the product, government officials removed the chemical from distributors' shelves and/or initiated legal action.[50] As this brief outline of the regulatory history of farm chemicals indicates, farm families were much more exposed to the dangers of agricultural chemicals than was the general public dependent on the nation's food supply.

Alice Hamilton contributed greatly to the emerging field of occupational medicine while devoting much of her professional energies to the lead poisoning issue from 1908 well into the 1930s. Her pioneering toxicology work demonstrated that lead exposure in factories received attention much earlier than did chemical safety on family farms. Hamilton's activities included numerous first-hand factory investigations, pioneering poisoning studies, and promotion of European approaches to industrial safety in the United States. American industry woefully lacked data, regulations, and professional expertise prior to the work of such safety researchers and activists. Thus, as the First World War approached, measuring dangerous lead levels, instituting lead dust reduction methods, as well as the overall awareness, had improved markedly. By the mid-1920s, Hamilton argued, the potential dangers of exposure to lead in the workplace had been reduced, despite the significant work that remained to be done.[51] The effort to reduce lead exposure in factories occurred fully two decades before a national farm safety movement had materialized. Historians such as Christopher Sellers have also definitively demonstrated how, by the 1940s, industrial hygienists, chemists, and a variety of other professionals had asserted their influence to improve worker safety across a variety of industries. Agriculture was not one of them.[52]

Despite these belated beginnings, the farm safety movement persisted throughout the century. The story of the organization Farm Safety for Just Kids provides a powerful example of the issue's longevity. In 1986, Marilyn Adams's son, Keith, accidentally fell into a grain wagon and was suffocated beneath thousands of pounds of grain.[53] Adams was overtaken by despair but eventually was able to transform her grief into a crusade to raise safety awareness by establishing Farm Safety for Just Kids. By 2003, Adams and thousands of others had created a national association with an annual budget of approximately $800,000 and annual membership reaching around fifty thousand children. Although the organization disbanded in 2016, it published a wide variety of safety materials, continually evaluated its efforts, and, through creative educational programs conducted by local chapters, instructed farm youth throughout the nation about the importance of safety. This important organization represented one recent development in a farm safety movement that has existed for more than sixty years. The effort signified the cooperative atmosphere that has been part of farm safety education since the Second World War.[54]

Farm Safety for Just Kids shares many tactics with other nonprofit organizations' previous attempts to prevent farm accidents. Such groups include the National Safety Council, 4-H, Future Farmers of America, and many other groups cooperating in mounting vigorous educational efforts to inform farm families about the many dangers of agricultural life. Farm safety activists enthusiastically work to improve farm families' living and working conditions. Similar to Farm Safety for Just Kids, predecessor organizations depended on private sector financial contributions to fund their efforts. Thus, Farm Safety for Just Kids symbolizes how voluntary educational efforts continue to play a crucial role in meeting the challenges of improving farm safety through continually evolving movements within a context of ongoing technological, political, and cultural change.[55]

2

A HOME AND A WORKPLACE

AMONG THE DIFFERENCES in living environments experienced in the mid-twentieth century by Corn Belt farm families and by urban families, one central aspect had the most far-reaching consequences: the farmstead functioned as both a home and a workplace. Teresa Klein, who grew up on a farm near Oelwein, Iowa, vividly described this situation stating, "People don't ever think of having their kids live at the factory with them, but that's essentially what you're doing; you're living on a very dangerous factory."[1] As a result, farm families encountered a dizzying array of hazards affecting rural individuals regardless of age, with both young and old exposed to potentially dangerous situations. Many of these were unique accident agents, such as those involving livestock, affecting only farm residents. Farm families of the era were coping with tremendous change as horse farming was eclipsed by industrialization. These technological changes included the adoption of new machines, storage practices, and chemicals.

Although both rural and urban residents sometimes encountered dangers stemming from the same sources, such as falls, destructive weather, and fires, these long-standing dangers were shaped, and even exacerbated, in the rural environment. Farmers, not urbanites, were exposed to possible tumbles from wagons or falls into the moving parts of an agricultural machine. Barns presented a more diverse variety of fire hazards than did city dwellings. In addition, farm families' outdoor lifestyles and relative isolation also influenced weather-related and recreational dangers, such as swimming pool accidents. Even the introduction of technologies common to city and farm alike, such as electricity, took particular shape in the rural context. Farm families used electricity for home conveniences and for a variety of work-related uses, often without the professional installation and supervision provided in other employment settings. The isolated nature of the homestead, too, affected the timely arrival of assistance in an emergency. In 1954, Raymond C. Johnson, a prominent 4-H leader, elaborated on this point, stating that fire prevention and reaction efforts were hampered in the countryside by the dispersed population and the impracticality of installing water hydrants.[2]

The farmstead's dual function as a home and workplace posed particular, unique dangers to the very young and the aged. Unlike in urban areas, in rural locales both demographic groups continued to be exposed to a highly hazardous working environment. In 1946, Maynard H. Coe, director of the National Safety Council's farm division, recognized this point.[3] The elderly were susceptible to accidents because, unlike workers in other settings, many farmers never fully retired: they continued to encounter agricultural hazards even as they became less able to adapt to them. Children were also exposed to many safety threats as they worked and played on the farmstead. The nation's agricultural leadership repeatedly acknowledged this dimension of the farm safety problem. In 1959, Ezra Taft Benson, the United States secretary of agriculture, encouraged the National Safety Council (NSC) to continue to focus its farm safety programs toward the farm family.[4]

In 1954, Rosemary Schaefer, a home economist for the NSC's farm division, offered a heartbreaking example illustrating the complexities of balancing the demands of home and farm. She provided the case of a woman whose spouse was forced to go on a trip at a time when much fieldwork needed to be completed. The woman was excited by the prospect of plowing a field that her husband had wanted to till prior to his departure, even though she had never driven the tractor by herself. Having no one to watch her four-year-old boy, she took the child with her. While attempting to till the field, she misjudged the severity of a steep hillside and overturned the tractor. When her husband returned home and looked for his family he found them lifeless under the tractor's crushing weight.[5] This incident powerfully demonstrates how agricultural families worked together and were consequently often exposed to unsafe situations. Even farm families seeking to follow the safest practice were often up against limits of time, labor availability, and experience.

Both the diversity and severity of childhood farm accidents suggest that parents were sometimes unable to keep their children safe. Youths met with accidents from exposure to a wide variety of hazards. Fatalities and injuries could result from farm machinery, electric fences, farm animals, and even from blasting caps used for stump removal. The farm child's access to blasting caps provides a revealing example of the challenges parents faced in trying to keep their kids safe. In 1944, a *Farm Safety Review* writer stated, "Children hammer or pick at them, throw them in bonfires or against walls, and the resulting explosion throws particles of metal in all directions. Fingers are blown off, hands are often crippled, [and] eyes destroyed."[6] Blasting caps presented a serious but rather uncommon hazard, but they nonetheless indicate the great diversity of farm dangers. Agricultural literature contains numerous examples of children who were injured or killed in more common types of accidents, such as those involving machines. In 1955, Harold Heldreth, an NSC writer, shared research on farm accidents in 1952: of the forty-five youths injured in mechanical accidents, thirty-five of them were involved in tractor mishaps. Heldreth also noted that the greatest proportion of fatal accidents involved children between the ages of one and four, with youths between the ages of fifteen and nineteen close behind.[7]

Little legislation had been introduced to protect vulnerable farm children. This was especially the case for those who worked on their own parents' farms. The laws enacted during the period prohibited youths under the age of sixteen from engaging in highly hazardous activities, such as operating agricultural equipment and applying pesticides or fertilizers. Unfortunately, these legal safeguards applied only to individuals who worked as farm employees; they did not protect the thousands of children who worked on their own families' farms.[8] Parents were ultimately responsible for the children's welfare and for assigning them age-appropriate work duties. In 1955, Heldreth echoed this reality, emphasizing that parents were accountable while acknowledging the incredible difficulty of the task, and illustrating the point with the tragic story told above of the woman and child killed while plowing in the husband's absence.[9] Ray Fuson, a high-ranking official for the Indiana Farm Bureau, also recognized how farmwork's stressful demands could easily expose youths to dangerous situations. In a 1963 statement he noted, "It is real easy for Dad to pull a 15-year-old boy off a $6,000 self-propelled combine, put him into a $175 truck and send him a few miles into town for supplies."[10] Clearly, some safety advice could at times seem highly impractical, given how much Corn Belt families depended on children as part of their agricultural workforce.

Farm boys and girls began work at an early age and were given jobs of increasing complexity and responsibility over time. Jim Klein, raised on a farm near Ashton, Iowa, in the 1960s and 1970s, stressed this point about his own upbringing, commenting that his "chore responsibilities kind of grew" as he got older, from tending livestock as a child to running equipment by the time he was an adolescent. Klein noted, "As long as you're old enough to hold a five gallon bucket of water, you can go and do your part on the farm."[11] Farm children clearly depended on their parents for safety since they governed their children's work tasks,[12] but equally clearly children made important work contributions to midwestern family farming operations at a time when child labor in industrial settings had been dramatically reduced under federal regulations. Even labor divisions by gender sometimes became fluid under the pressure of the need for labor. Historian Pamela Riney-Kehrberg has noted that although boys "typically worked beside their fathers, it was not unthinkable that a boy might cross gender lines and participate in chores that were common to girls."[13] Farm girls also sometimes could be found doing outdoor work that might have typically been viewed as male tasks.[14] Generalizations about family farmwork are thus made difficult, since the size, gender, and age of the child as well as the unique economic circumstance of the family all affected the distribution of responsibility for individual farm duties.

The experiences of other farmers confirm Klein's observations that, regardless of family circumstance, children were usually given work responsibilities at young ages. Leroy Hogeland, a lifelong farmer from Monroe County, Iowa, stated that he began doing chores "as soon as my hands were big enough to hold a hoe."[15] James Wilson, another southern Iowa farmer, remembered serving as "water boy" for a threshing crew on his grandfather's farm in the 1930s. Driving a pony cart that held five stone jugs filled with cool water, he traveled back and forth from the water source to the threshing crews in the field. Although Wilson described the work as tiring and sometimes found his tasks frustrating, he rarely felt overwhelmed by his work as a child.[16] Rose Mary Scott, who has lived on Monroe County, Iowa, farms her entire life, vividly remembered the many tasks she did as a youth, including "getting the cows to milk, gathering the eggs, [and] working in the garden." Rose especially recalled the tedious task of cleaning the cream separator as one of the many ways she contributed to her family's farm operation.[17]

Children encountered hazards in numerous and unexpected ways on the nation's farms and were often exposed to those risks at an early age. Farmer Bernard Stodghill stressed that he was made aware of the farm's dangers as a small child when he witnessed a horse's tail get torn off after it was caught in a threshing machine pulley. His dad used a "rubber jar ring" to stop the bleeding and kept the horse from panicking, but the incident remained a vivid reminder for Stodghill of the farm's many hazards.[18] Such incidents not only reveal one of the many ways children could get hurt on the farm, but also how hard it would have been to ensure a child's safety when they were so integrated into the life and work of their farms.

A child's natural inexperience and curiosity combined with the farmstead's complexity posed special problems. In 1955, a *Farm Safety Review* writer stressed that a child's natural inquisitiveness needed to be tempered with safety awareness. Another safety advocate expanded on the point stating, "Consider your children. Does your toddler go unwatched as he explores his constantly expanding world? He needs your constant attention except when he is in a play pen or fenced area where his activities are limited."[19] In 1954, Rosemary Schaefer, an NSC home economist, offered safety advice that illustrates the challenge of keeping farm kids safe. She suggested the implementation of practices such as establishing specific play areas, giving children age-appropriate work, and consistently instructing kids about safety.[20] Although such precautions would appear rational to contemporary readers living in urban areas, such practices were undoubtedly impractical to implement on family farms. The idea that parents should give "constant attention" to their children's safety might have represented an ideal scenario, but it ignored the reality that farm families lived in their workplace twenty-four hours per day.

Children's innate inquisitiveness made them especially susceptible to livestock mishaps. Schaefer emphasized that parents should supervise all interactions their children had with large animals such as horses.[21] Heldreth also emphasized the special concerns related to children and animals, particularly horses, by writing "thousands of farm boys and girls are in the saddle daily to carry out certain farm chores more quickly. Many young people ride horses for recreation."[22]

Farm youth were exposed to dangers associated with horses and other farm animals at an early age. Agriculture presented risks long before the advent of tractors, corn pickers, and other machines that came to dominate Corn Belt agriculture in the mid-twentieth century. *A.M. "Pete" Wettach Collection, State Historical Society of Iowa, Iowa City. Reproduced with permission.*

He suggested that interactions with animals should be supervised, age appropriate, and that all farm youth should be trained in proper livestock handling.[23] Again, such advice represented the ideal scenario, given that parents faced challenging work schedules that would have made such practices difficult or impossible to implement.

Despite their parents' best efforts to prevent accidents, farm children were sometimes hurt by horses. Lawrence Schmitz, a farmer from Wapello County, Iowa, provided a vivid example of the dangers horses presented for children. He said that one of his friends "got too close behind a horse and it kicked him and broke his jaw." Schmitz reflected on the event decades later and emphasized that the injury had had a permanent impact on his friend. Although he had received medical treatment, his friend's jaw "was never the same." As a young man this friend worked for the Schmitz family, and Lawrence Schmitz related that "if we had sandwiches in the field or something, he'd have to smash the sandwich down" to eat it.[24]

Hogeland also related a painful experience he had while working in the barn at the age of nine. He received a painful bite from a particularly "ornery" horse. He stated, "She reached around and got me by the shoulder and actually picked me up." The horse attack left a nasty wound and a lingering distrust of horses. Hogeland summarized his experience with horses as a youth by stating, "If I had to have been a horseman, I would have had to have been something else; I would not have been a farmer. I was never a horse lover that much. I had some really pretty bad experiences, like biting. Dad was kicked and thrown off of horses."[25] Although some farmers had pleasant memories of horses, such testimonies reveal that horse accidents were common among farm children.

Exposure to potentially dangerous technologies combined with parental negligence contributed to childhood fatalities. Tractors were a frequent cause. Children were most usually killed as a result of being a second rider and getting caught in a tractor's power take-off shaft or when playing around a moving tractor. In 1962, FFA members of Ellsworth, Kansas, spoke at the National Safety Congress, stressing how unwise it was to give children tractor rides. The speakers indicated that a farmer who allowed a second rider "doesn't value the life of his children very much, or he wouldn't let them ride on the tractor." Although such comments may have been harsh, having a second rider was indeed very hazardous. A small bump in the road could throw a child from a tractor and result in serious injuries.[26]

Yet the worth of second-rider warnings must be balanced against an understanding of how farmers learned the finer points of their profession. Farmers justified the practice despite many safety experts' warnings against having a second rider on a tractor. A key aspect of family farmwork was the training process, which relates directly to the practice of allowing a second rider. Unlike workers in a manufacturing setting who might receive special training, farmers usually relied on observation of family members to acquire much of their agricultural knowledge, including machinery operation. They often did so at very early ages. Craig Fausch, a farmer from Story County, Iowa, emphasized the point by stressing the many hours he spent as a second rider on his dad's tractor. He indicated that he was with his father every day and that he "lived and breathed" farming from an early age.[27] Don Rimathe, also of Story County, Iowa, stated, "You can learn about the science of farming, but to learn how to do the job—I

Children often accompanied their parents during farm work. Many farmers stressed that they had learned much of what they knew about agriculture by working alongside their parents. Adults also viewed such situations as opportunities to convey knowledge to their children. Unfortunately, such methods of learning sometimes exposed children to significant risk. *A.M "Pete" Wettach Collection, State Historical Society of Iowa, Iowa City. Reproduced with permission.*

just don't think you can learn it without being there."[28] These insights reveal how young people gained their knowledge of farming but also how early in life they were exposed to farm hazards.

Other influences on the work roles and even safety of farm children were birth order and family size. Jim Klein noted that, since he had an older brother, he assumed farmwork chores at a later age than his brother had. He stated, "My brother was four years older than me, so he always had a larger share of the work than I did." Klein added that as the years passed, work could be divided among all four brothers, but initially his brother "had to do it all by himself."[29] James Wilson, who had two brothers, revealed that sibling relationships affected work and safety conditions in other ways. He stated, "I was the oldest, and I was trying to get my brothers to do things that I thought they ought to be able to do, but in retrospect they were too young."[30]

In addition to hazards faced by children, safety writers pointed to the growing problem farm dangers posed to the elderly. Writing in 1953, Harold Heldreth stressed that many farmers, due to the deeply personal connection they felt to their occupation, never really retired from farming. Heldreth also pointed out that the nation's agricultural production would be increasingly dependent on older workers, commenting that "today, 50 percent of our workers are over thirty years old and 22 percent are over fifty years old." The elderly were particularly vulnerable to farming-related injuries, however, due to their decreasing physical abilities and much slower recovery from injuries. Heldreth indicated certain jobs and practices that venerable farmworkers should avoid: among them, overwork, use of equipment under difficult conditions, and highly strenuous activities. Despite the prudence of Heldreth's advice, it is easy to imagine that many farmers might ignore it.[31]

FATIGUE AND FARMING

Farmers were also exposed to risks for long periods of time, resulting in fatigue. In 1946, a *Farm Safety Review* writer recognized this tendency, highlighting farmers' propensity to work to the point of decreased alertness, noting that "on the farm, that is especially bad, for the farmer lives on his place of work, so is exposed to hazards of his occupation for longer periods than most workers—even in normal times for twelve hours a day and seven days a week."[32] James Wilson also indicated that farming was particularly stressful during harvest time. He remembered several times falling asleep while harvesting crops late into the night.[33]

The traditional demands of the harvest weren't the only aspects of occupational stress contributing to farmers' vulnerability to accidents; economic trends were at work as well. Many farmers fulfilled their agricultural responsibilities while holding second jobs. Bernard Stodghill was a lifelong farmer who at the same time held a full-time job at Anderson Erickson Dairy in Des Moines, Iowa.[34] His dual work responsibilities were not unusual, especially for farmers who owned or rented small acreages. In 1969, D. B. Pointer, a farm editor for the *Jefferson City Post-Tribune*, reflected on this trend by stating that "the high cost of living, increased taxes, and a dwindling market price for farm products in relation to land and equipment costs, has caused a

These farmers are fortunate to be baling together, since farmers sometimes found themselves alone when serious accidents happened. Sadly, farmers were sometimes found dead hours after a lethal accident. *A.M "Pete" Wettach Collection, State Historical Society of Iowa, Iowa City. Reproduced with permission.*

stampede of small farmers seeking a job in the city to supplement farm income."[35] It is difficult to know the degree to which this type of stress contributed to accidents; however, in the early 1960s, Stodghill sustained one of his worst injuries after a shift at the dairy. He caught his hand in a tractor-mounted corn picker and had to drive himself to a neighbor's house for emergency assistance. Although he recovered, Stodghill lost part of a finger, a permanent reminder of his accident.[36]

DANGEROUS ANIMALS

Farm families were also much more exposed to possible dangers from animals than were other populations in the country. These went beyond the possibilities of dog attacks and other pet-related injuries possible in cities. In 1945, Marvin Nichol, an NSC farm safety expert, indicated the significant magnitude of the problem when he noted that incidents involving animals were surpassed only by those from farm equipment as causes of accidental agricultural deaths.[37] Horses could inflict injury by biting, bolting, kicking, or stomping. Cattle did not bite, but they inflicted injury when they kicked with their hind legs, gored with their horns, or stepped on their owners' feet. Pigs could inflict wounds with their tusks, and a male sheep could ram with its horns. Farmers even encountered hazards while doing routine tasks such as milking. Rex Peterson, a lifelong Appanoose County farmer, remembered one cantankerous milk cow that could "kick to kill you."[38] He used a device known as a kicker to prevent the cow from hurting him.

Children as well as adults were exposed to injury from farm animals and learned about such potential hazards at an early age. Perry Middlesworth, a lifelong farmer from Wapello County, Iowa, learned to respect cows at an early age and to be particularly cautious at calving time. He told how as a boy he was once checking on the cows and got too close to a calf. Its mother "knocked me down and I ran and jumped over a fence. I learned a big lesson; you don't bother a mother and her baby."[39] Unfortunately, some farm kids had even closer calls. Jim Klein related a story in which he and his brother were sorting cattle when a steer bolted toward them. Klein fell down as the large steer jumped over him. He remembered seeing the steer's belly as the animal sailed above him. He jokingly recalled thinking, "I hope he makes a long jump." Klein also recollected his father stating, "You probably don't have to tell mom about this."[40] Such comments confirm the ways in which parental concerns for safety might be subsumed under the need for labor.

Leroy Hogeland raised hogs on his farm for decades, and he emphasized the unique hazards associated with raising twelve to fourteen hundred pigs a year. He stated that moving hogs was exhausting because of their weight and because they had minds of their own. Boars required special attention: "They would have tusks on the side of their jaws that would be razor sharp and could cause one hell of a wound." Hogeland indicated that in the 1960s and 1970s he used a head-gate and a hoof trimmer to "de-tusk" them. He stressed that de-tusking was a safety measure but that it was important to conduct the procedure on all his boars. He remembered that on one occasion a male pig sensed that his rival didn't have tusks and "tore the other one all up."[41] Rex Peterson also noted that a bad-tempered boar could inflict a serious wound and required caution.[42] Such observations emphasize the experiential nature of agricultural knowledge as it related to safety both for people and for their farm animals.

Bulls in particular demanded special respect due to their large size, strength, instinctual protectiveness, and aggression. They also possess horns capable of inflicting serious injury. Safety advocates at the time warned that bulls could be one of the most dangerous of all farm animals since they could cause wounds by "goring, butting, hooking with horns, and trampling."[43] In 1943, E. S. Bayard, editor of the *Pennsylvania Farmer*, wrote about the significant scope of the problem based on his then twenty-year study. He had recorded 579 bull attacks, of which 214 resulted in death. Bayard emphasized that many bull injuries were never disclosed and that sometimes even local newspapers did not cover them, indicating that the issue was probably more serious than most believed.[44] In 1944, Randall Swanson, a Wisconsin farm safety specialist, revealed that in the first half of that year alone, six farmers in his home state were killed and another twenty-five severely injured in accidents involving bulls.[45]

John Van Ringelestein vividly described his frequent encounters in the 1950s with a particularly unpredictable and bad-tempered bull, which had on more than one occasion chased him. Eventually his father solved the problem by getting rid of the dangerous animal.[46] Rex Peterson recalled one incident when he got in the middle of a fight between several bulls. One of the bulls involved, Peterson stated, "was an ornery little devil. He didn't bother me, but he didn't like the other bulls. Eventually, he got into a brawl with another bull while I wasn't looking. They hit me and knocked me head over heels. I got some broken ribs and twisted glasses and I think I was out for a little while."[47] Bulls were clearly unpredictable and deserved great respect. Even seemingly docile bulls could present significant danger, especially since farmers were sometimes less concerned while handling them.

Farm accident specialists suggested that building "safety bull pens" was one of the more effective methods of preventing such violent encounters.[48] A safety bull pen usually included a shelter, a breeding stall, and a large exercise enclosure surrounded by a sturdy, high fence. This bull-handling system was highly beneficial to the entire family's safety since it greatly reduced any contact with a bull. In 1944, C. L. Hamilton, a prominent agricultural engineer, expressed this point, stating that the enclosure allowed farmers to conduct many necessary interactions with the bull while lessening vulnerability to attack. In 1944, Randall Swanson stated, "We can't talk too much about the safety bull pen since we still have to recognize that 95 percent of Wisconsin farmers do not have these facilities for handling bulls."[49] The fact that farmers may not have adopted such livestock handling procedures quickly enough for Swanson demonstrates that farmers did not always view changes to their practices in the same light. Farmers implemented such preventative measures within the time and budgetary constraints of their own situations. Farmers possessed limited financial resources and lived within the context of increasing production costs. Farm safety experts, on the other hand, although devoting their careers exclusively to farm safety, were paid largely by government agencies, which goes a long way in explaining such diverging approaches to farm safety priorities.

Pat Larkin, a lifelong beef cattle producer from Wapello County, Iowa, recognized the importance of improvements in livestock handling systems. He stressed that such developments were crucial in lessening the hazards associated with working cattle. Larkin worked for a ten-year period at the Iowa State University Extension Service and assisted professors in improving livestock structures. He stated, "I got into designing a lot of facilities for working cattle and we pretty well got it designed where you never got in a pen with the cattle." He said, "We used to grab a calf and throw him down and vaccinate him, but now we drive him up an alley way, reach through, and poke him a little bit with the use of a head gate." He underscored that such structures reduced hazards for people and lessened animal stress while greatly increasing cattle-processing speed. Larkin also highlighted that such systems emerged not only from practical experience but also from animal behavior research.[50] Farmers' exposure to cow-related hazards was also reduced by other innovations, such as artificial insemination, widespread by the mid-1950s.[51]

Farm families' greater contact with livestock also potentially exposed them to diseases that could be transferred from animals to humans, especially rather common maladies such as rabies and brucellosis (Bang's disease). Rabies presented the most significant threat. Although cattle and cats were capable of spreading the sickness, dog bites were the most common transmission method. Brucellosis, which also garnered attention, could be transferred to humans through consumption of infected milk, from an open wound, or through the air.

People suffering from the disease might experience fever, fatigue, and weight loss.[52] W. A. Aitken, a veterinarian and member of the American Veterinary Association, emphasized that agriculturalists were faced with higher risks of contracting such diseases because of their more frequent exposure to animals. He also indicated that farm families faced health risks from as many as nine diseases commonly present on farmsteads. Safety writers advised precautions such as having livestock tested, wearing rubber gloves while handling animals, and thoroughly cooking meat.[53]

Health officials also warned farm families about less common illnesses, including leptospirosis, trichinosis, and tularemia. Leptospirosis affected a variety of animals, including cattle, pigs, and even dogs, and could be transferred through diverse means, including soil, urine, and water. An individual might become infected by swimming in a pond contaminated with infected animal wastes. Symptoms include yellow skin, fever, and diarrhea. People usually caught trichinosis by consuming insufficiently cooked and infected pork. The animals themselves usually acquired the disease from the unwise practice of feeding garbage to pigs. The parasite that caused the disease resided in the infected animal's tissues and entered the human digestive system upon consumption. Trichinosis caused symptoms such as severe stomachache, fever, and coughing. Tularemia, a bacterial infection, also resulted in symptoms such as stomachache, fever, and cough. Wild game, not domestic pigs, were usually the disease source, contracted from consumption of infected animals or insufficiently cooked meat. In 1965, a *Farm Safety Review* contributor emphasized that infected rabbits posed a particularly high risk of transmitting the disease.[54]

FIRE DANGERS

Although all Americans faced fire hazards, farm families encountered additional fire risks that were distinctive to the farm and reflected the ongoing process of farm industrialization. In 1944, Burton Williams, a *Farm Safety Review* contributor, highlighted the unique hazards associated with barns, including improper straw and manure storage, which could lead to spontaneous combustion. Williams also stressed that storage and use of ever-growing amounts of petroleum products presented dangers, making the careless disposal of cigarette butts or matches especially hazardous. Burton's identification of petroleum products as the causes of many fires reveals farmers' increasing dependence on such products for fuel and machinery lubrication. Farmers could easily multiply the hazards by not storing gasoline in suitable containers with flexible spouts, flash arresting screens, and seal-tight caps. In 1949, A. M. Sowder, chairman of the USDA's Fire Prevention Committee, also warned farmers to properly dispose of "oily rags," which could easily ignite. Safety advocates advised farmers to place oily rags in "self-closing metal containers" that used foot pedals to open and then closed automatically, instead of leaving them strewn on floors.[55]

Farm families also encountered potential fire hazards from the growing repertoire of heating equipment. Raymond C. Johnson, a prominent 4-H leader, revealed the dangers associated with heating equipment, including grain dryers used to treat moist crops and heat lamps used to keep animals warm. If used improperly, such devices could easily ignite the highly combustible materials typical on farms. Midwestern farmers had to be particularly aware of such risks when drying their grain. In 1962, a *Farm Safety Review* writer suggested that a farmer's grain dryer should have a "thermostatic control to shut off the blower or the dampers when the temperature in the heat transfer chamber gets too high." The author also warned that the dryer should be vented to ensure that the machine's exhaust was emitted outside.[56] Despite the wisdom of such precautions, farmers undoubtedly varied greatly in their understanding of heating and drying equipment and would have had varying degrees of skill in installing and maintaining such devices properly.

Farmers also needed to operate their machinery with fire prevention awareness. Observers suggested that fire extinguishers be within easy reach when operating tractors, corn pickers, and combines. In the mid-1960s, David T. McFarland, an NSC staff member, advised farmers to always turn off their equipment prior to replenishing their gas tanks, to keep the components of a hot engine clear of debris, and to fix leaks promptly. In the early 1970s, Joseph Gerling, a farm safety specialist employed at Oklahoma State University, warned farmers that a combine, a desiccated field, and hot weather could create dangerous conditions. He advised farmers to "periodically check for overheated bearings that could burn or start a fire in dry chaff. Be particularly careful to keep the exhaust manifold and other parts of the exhaust system clean and free of dirt and chaff. Be alert for constantly slipping belts which could heat up and catch fire."[57] Although such advice was useful, farmers most likely found such practices difficult to follow amidst the pressures associated with the Corn Belt harvest season. Illinois farmer Robert Westpfahl commented that some were willing to work incredibly long hours during the harvest. He stated that "[s]ome farmers just run these machines into the ground and there is no machine safer than the man running it. If a man has been running the machine for sixteen or eighteen hours a day for seven days a week, he is getting dangerous."[58]

Wesley Longnecker recalled having firsthand experience with fire hazards related to design features of his corn picker. He remembered that in the early 1940s, while harvesting corn with a two-row corn picker mounted on an M Farmall tractor, his machine caught fire on several occasions. He said that "there was a glass sediment bulb underneath the gas tank and that sometimes it would catch on fire right in the center where you were sitting." He also stressed that it was important to put those fires out quickly or it could escalate into a very dangerous situation. Longnecker's comments reinforce the importance of constant attentiveness in avoiding fire accidents as in avoiding many other hazards associated with farmwork.[59] Rex Peterson also found that older equipment represented unique hazards associated with heat and fire. Peterson recalled that his dad had a Fordson tractor and that "there was a worm gear in the final drive and it would just about cook you; it was just about like sitting on an oven."[60] Safety advice could thus be undermined by farm equipment engineering problems. Such design issues were exacerbated by the time and financial pressures associated with farming in the mid-twentieth century, which often required long hours with few breaks.

ELECTRIFICATION

During this period, some of the same technologies were transforming both rural and urban life, although under widely varying conditions of installation, use, and maintenance, and this included adoption of electricity. In 1938, less than half of the nation's farms were electrified; however, over the following two decades, a majority of rural families adopted it. James Wilson, who was born in 1927, spent his entire youth on a farm that did not fully benefit from electricity.

Nevertheless, he vividly remembered his father's excitement at the installation of a single light bulb in the home. He recalled that once, when his mom was away on a trip, his dad mounted a bulb powered by a "Delco Power Plant" in the "center of the house." Wilson said that his dad was "so proud of this because we had been using kerosene lamps. He reached up and turned the light on, but mother came in and didn't even notice." She had been in town where electrical lighting was common. Wilson thus reveals both the excitement of many rural Americans at the prospect of electricity and the gulf in standards of living between urban and rural America.[61]

Other new conveniences included hot running water, refrigeration, and a variety of other amenities. During the 1950s, the greater availability of telephones also shrank the technological gulf between the city and the countryside. In 1962, Judy Styles, a member of a South Dakota 4-H, celebrated the consequences of the electrical access stating that "[i]f we were living in the days when our grandparents were our age, we probably wouldn't be here today. We might be possibly taking our turn at the family washboard or some other laborsome device and wouldn't have so much time to do the things we enjoy. What made this possible? Electricity!"[62] In 1964, D. E. Lindstrom, a rural sociology professor from the University of Illinois, echoed such remarks, noting that most farm families enjoyed "a good modern home with electricity, hot and cold running water, television, radio, telephone and many of the labor-saving devices such as a sewing machine, refrigerator, deep freeze, and other home conveniences."[63]

Electricity improved safety, since better lighting helped farmers navigate their crowded environments. In 1944, Myrtle Fahsbender, director of residential lighting for Westinghouse, credited electrical lighting for reducing instances of tripping, falling, or running into unseen objects. She encouraged farmers to adopt electrical lighting quickly since it improved visibility during periods of low light, such as evening or early morning. In 1952, Earl L. Arnold, of the Rural Electrification Administration, also believed that "a well-lighted farm had reduced common night time and early morning accidents such as banging one's shin into farm equipment, twisting ankles in holes, and a host of other mishaps partially caused by darkness."[64] In addition, electric lighting was probably safer than earlier sources of illumination.[65]

Despite such benefits, careless electrical practices still produced serious consequences. In a speech given at the 1962 National Safety

Conference, South Dakota 4-H club member Judy Styles cautioned her listeners against irresponsible habits. She told the story of a rural Tacoma, Washington, couple who lost seven children in a blaze caused by an over-heated refrigerator motor. Styles stressed that electrical mishaps had become a leading cause of house fires. In 1975, William J. Fletcher, a farm safety specialist, also emphasized the dangers associated with careless electrical use, stating that "[e]lectricity has become one of our most essential services to our farms and ranches, but few things we use from day to day in our homes and in agricultural work have greater potential to do us harm than electricity." He also indicated that approximately one hundred farmstead electrocutions occurred annually on the nation's farms and that electrical mishaps were the principal cause of fires.[66]

Such comments illustrate the larger point that farm families assumed a variety of maintenance and home improvement tasks that some experts may have considered risky. The technical expertise to run a twentieth-century farm could surpass the knowledge base of even the most progressive farmer. In addition, the rural setting influenced the nature of electrical dangers. Because farms represented both residences and workplaces, rural electrical users employed electricity in more diverse ways and often in greater amounts. Farm families lit not only their homes but many other buildings, including barns, livestock sheds, and storage facilities. In 1956, William Peterson, a South Dakota Extension electrification specialist, recognized the farmer's ever-growing electrical dependence. He stated that "[e]very winter, traditionally in December, the Great Plains farmer hooks on a record electrical load to his farm wiring system. This extra load increases from 10 to 20 percent each year." Peterson's observation confirms farm use of electricity for lighting, heating, pumping water, and a host of other purposes.[67]

Observers also noted that increasing electrical use could strain outdated wiring systems. Judy Styles stated that "our real trouble is not with short circuits but with overloaded circuits. The basic problem is this: We have quadrupled our home consumption of electricity in the last twenty-five years, largely on wiring designed to handle 1933's load." She continued, "A generation ago, there were only about fifteen types of electrical appliances in use. Now there are more than fifty."[68] In 1956, Peterson recognized this problem and recommended that farmers update their electrical systems just as they replaced worn-out and obsolete agricultural equipment. Another *Farm Safety Review* author commented that "older installations often are not up to present power demands. Wires get hot circuits and are over loaded. Age and weathering increase the possibility of failure."[69] Even as late as 1975, William Fletcher was stressing the problem of outmoded electrical systems, revealing how persistent the problem continued to be. Farmers faced dangers not only from overloaded systems, but from the expanding repertoire of devices that utilized the new power source, including heat lamps, electrical fences, and even the electric lines delivering power to their farmsteads.[70]

Both the heat lamp and electrical fence could be dangerous if used or installed improperly. The heat lamp, unlike traditional incandescent lights, provided the benefit of "radiant heat rays," used to keep brooding chickens warm, to prevent drinking water from freezing, and to dry newly birthed livestock. However, careless use could result in destructive fires that could kill livestock and destroy buildings. In 1954, a *Farm Safety Review* writer stressed such proper installation practices as using moisture-resistant cords, securely suspending the lights, and placing the lights at a proper distance from the animal bedding. An improperly installed electric fence was also hazardous for adults, children, and even livestock. C. L. Hamilton cautioned farmers against using "home-made installations, direct hook-ups, make-shift current regulators, or unsafe commercial products when installing electric fences."

Hamilton also warned that electric fences could present dangers if the fence was placed "near stock tanks, ponds, irrigation ditches or damp ground." He provided a particularly heart-breaking example illustrating this point: "In the fall of 1944, the merriment of three little boys sailing sticks and splashing water from a farm stock tank in Northern Illinois was suddenly stopped when one of them was electrocuted by an unsafe electric fence which ran by the tank." He added that "this fence was connected directly to a 115 volt lighting circuit with nothing but an ordinary fuse to limit the current." Electric fences were also involved in livestock deaths, as supported by the experience of employees at a single Corn Belt rendering plant that annually collected over forty farm animals killed from fatal electric-fence encounters.[71] Teresa Klein remembered a painful episode when she was a preschooler. She "got [her] hair caught in an electric fence" while attempting to crawl under the wires. Her excruciating encounter taught her to "never touch an electric fence again."[72]

Farmers not only faced safety challenges from the various electrical devices they used, they also used them under conditions that differed greatly from urban settings. In 1975, Fletcher remarked that "corrosion from manure in livestock pens will make quick work of thin wall galvanized conduit. In such areas coating the conduit or otherwise protecting it from manure is a must. Plastic conduit might be used if not subjected to extra physical abuse." He also stressed that a spark originating from a switch or a broken light bulb might be benign in most settings, but it could prove disastrous on the farm. He advised that safety devices such as "explosion proof switches" and "fixture guards" be installed, since farm buildings contained fuel, hay, and livestock feed, all of which were highly flammable; such devices kept sparks from reaching combustible materials and thus helped prevent fires.[73]

Electricity also posed dangers when users lacked knowledge of its basic properties. In 1949, a *Farm Safety Review* contributor emphasized just how little power was required to cause a fatal accident. The author stated, "Few people realize how small an electric current is needed to cause death. The ordinary 115-volt lighting circuit can be as deadly as 40,000 volts. It is current amperage that kills, not voltage."[74] Hamilton, an agricultural engineer, expressed alarm at the public's ignorance of the power source since its dissemination necessitated that people understand the basics of electrical safety. He provided examples illustrating hasty or unsafe behavior with electricity, including improperly installing electrical motors, placing appliances near water, and hitting highline wires with farm equipment.[75]

Farmers' electrical dependence also required that they perform regular electrical maintenance by repairing worn and frayed cords and other broken components to prevent fires. Safety advocates alerted farmers to a lengthy list of repairs that needed to be performed regularly. In 1968, Jack Burke indicated that such upkeep included finding and repairing "worn-out or malfunctioning equipment, electrical materials that have deteriorated from years of service or unusual stresses (flexing, weathering etc.), short circuited wires or equipment, loose electrical connections, no grounds or poor grounds, overloaded motors, equipment, poor placement of equipment that create fire, shock, or tripping hazards."[76] Burke maintained that electrical components sometimes needed replacement since time and continued use increased the chances of a malfunction. The problem of farmers neglecting such responsibilities thus appears to have persisted throughout the era of our discussion.[77]

As early as the late 1940s, observers were realizing that the sophisticated nature of electric wiring might exceed the farmer's skill level; this concern became heightened as rural electrification expanded. Farmers were required to keep up electrical systems that were larger in scale and more complicated than those in residential areas. H. H. Beaty, an Iowa State College agricultural engineer, encouraged farmers to utilize professionals, stating that "wiring installed by trained electricians helps prevent fires." He also recommended using products that conformed to industry standards, while warning against the use of "homemade or unimproved transformers on electrified fences."[78] Almost a quarter century later, William J. Fletcher revealed that some farmers were still maintaining their own electrical systems. He cautioned farmers against allowing novices to perform maintenance and advised that qualified individuals be consulted. Regardless of income level, farmers hesitated to take such advice, since most valued saving money.

Farm families clearly assumed a variety of maintenance and home improvement tasks that some may have considered risky. John Van Ringelestein stated, "I put the electricity in both barns in my late teens." He learned these electrical skills largely by observing a professional electrician wire his family's home.[79] Teresa Klein indicated that her family also did most of their own electrical work. Her husband, Jim Klein, recalled that his older brother learned welding in high school and as a result was responsible for much of the family's welding needs.[80] Leroy Hogeland captured the can-do attitude of many farmers, stating that "you either did it on your own or you didn't get it done."[81] Although some farmers did rely on electricians and professionals, many embraced this self-reliant attitude that could potentially lead to trouble.

Some farm dwellers were willing to expose themselves unnecessarily to electrical dangers. In 1952, G. E. Henderson of the Southern Association of Agricultural and Vocational Engineering stated, "Most farmers have been quick to see that electricity saves labor and cuts production costs. It is much harder for them to realize that proper wiring also cuts costs and saves lives." He went on to indicate that farmers often did not possess sufficient knowledge to manage their own electrical needs and were overzealous in their attempts to save money at the expense of safety. He related a story of one Mississippi farmer who installed a pump by using a damaged lamp cord to attach a socket and expressed great satisfaction when the pump functioned without the need to consult a professional. Unfortunately, the farmer's solution presented serious safety hazards for both himself and his family.[82]

NATURAL HAZARDS

Farm families' outdoor lifestyle also exposed them to severe weather conditions, including oppressive summer heat, lightning, and freezing temperatures. A farm family's work had to be completed regardless of seasonal extremes, thus exposing them to weather-related illness such as heat stroke or hypothermia. Climate-related dangers such as tornadoes and lightning hazards were also shaped by the countryside's special physical conditions. In 1960, a *Farm Safety Review* writer stated that "sprawling farms, unprotected by high city buildings, are excellent lightning targets; the loss in farm property from lightning fires is estimated at $56 million."[83]

Ruth Falck, who grew up on a farm in northern Minnesota in the 1940s and 1950s, emphasized the challenges of living in her native state: "The coldest I've seen it is 55 below zero," but she noted that it would have been worse if it had been windy. She recalled that the kids "always had white eyebrows" by the time they got to school during the wintertime. Falck stressed that children had to be aware of dangers associated with wild animals, as well, since their outdoor lifestyle included "a lot of hunting and fishing." She also pointed out that in her part of the state, which was heavily forested, forest fires would sometimes occur. Although Falck's account is probably somewhat atypical, due to the especially severe nature of the Minnesota environment, her recollections nevertheless reveal the distinctiveness of living in the countryside.[84]

Although farm families often faced extreme weather, they were prone to ignore proper preventative weather-related measures, since completing farmwork took priority. In 1960, Purdue researchers stressed that farmers could avoid heat-related injury by slowly acclimating themselves to hot weather, taking short but frequent work breaks, and wearing loose-fitting, light-colored clothing. Farmers were also advised to adequately hydrate themselves. An NSC staff

The Corn Belt farm was frequently the setting of recreational pursuits, such as swimming, hunting, and fishing. If anyone was injured, assistance from neighbors and emergency services was distant. *A.M "Pete" Wettach Collection, State Historical Society of Iowa, Iowa City. Reproduced with permission.*

writer provided similar advice while adding that taking salt tablets compensated for the minerals lost through perspiration. Although such advice was prudent, overburdened farmers sometimes did not fully implement such precautions. Pat Larkin believes that the many years he worked unprotected out in the sun most likely contributed to the "skin cancers" he now suffers as an older man.[85] Jim Klein remembered the cold weather also presented challenges, especially on occasions when he and his brothers had to chase cattle that had escaped over the fences during periods of high snowfall. He recalled that they would have to really bundle up; nevertheless, one of his brothers did suffer some frostbite. Klein recalled that one of his brothers suffered enough damage that his beard is "blotchy," and they jokingly call it the "Atrazine beard."[86]

RECREATIONAL DANGERS

Farm families also faced a variety of recreational hazards, especially those associated with swimming. In 1959, a *Farm Safety Review* piece indicated that leisure pursuits were responsible for a significant number of rural fatalities; in fact, 14 percent of accidental fatalities were caused by drowning. The dangers associated with recreational water use not only included drowning but also boating accidents or injuries with fishing hooks. Rural Americans were exposed to water-related risks largely from the farm ponds dotting the countryside. Farmers used ponds for watering livestock, for irrigation, and as a fire-fighting resource, but they also provided recreational opportunities. The dangers of drowning represented the most serious safety risk of farm ponds, particularly among youth. In 1966, a study provided a composite picture regarding the various factors contributing to rural drowning accidents. Jack Burke commented on one of the most distinguishing features of these small bodies of water, emphasizing that, unlike city pools, no lifeguards were present and children were a long way from help. The investigation also confirmed that pond-drowning deaths were a leading cause of accidental youth fatalities.[87]

THUS, IT IS CLEAR that farm families existed in a complex and, in many ways, distinctive environment. Although such age-old dangers as those associated with livestock handling and accidental fires have always presented dangers to farm families, new technologies such as electricity deeply altered the nature of the risks of farm life. The farm's dual functionality as both a home and a workplace exposed people of diverse ages and skill levels to an entire host of potentially dangerous situations. The farm's relative isolation in comparison to urban residences and worksites also hampered the speed with which agricultural workers or rural accident victims could receive treatment. Such aspects of the farm safety problem both deeply shaped the farm safety educational movement and contributed to the persistence of hazardous conditions. Although the aforementioned dangers represented a significant challenge for farm safety advocates, the dramatic mid-twentieth-century revolution in farm machinery also deeply altered the nature of farm hazards.

3

MACHINERY HAZARDS

ALTHOUGH FARM EQUIPMENT DANGERS before the mid-twentieth century had attracted little publicity, families had been adapting to new agricultural machinery and their dangers long before the emergence of the farm safety movement. Colonial farm families had labored with technology that varied little from that used during antiquity. They depended on spades, hoes, sickles, and flails to plant, harvest, and process crops, just as had their ancestors. Yet, during the nineteenth century such farm implements were replaced with a variety of equipment, including new types of plows, mowing machines, and reapers, to meet the rising needs of an ever-growing population. Thus, while technological change in agriculture had formerly been slow, innovations dramatically accelerated from the nineteenth century onward.[1]

During the Civil War, Corn Belt farmers rushed to meet market demand and address labor scarcity by adopting new equipment, which intensified the rate of mechanical change. Farmers continued to transform their work throughout the late nineteenth century. One of the most significant changes involved farmers' adoption of steam-powered equipment. Farm equipment companies were producing over five thousand steam tractors per year by the twentieth century's first decade. Steam power in turn would be replaced by dependable gasoline-powered tractors, such as International Harvester's Farmall tractor, in the 1920s. Such changes collectively contributed to the transformation of agriculture from a largely organic to an industrialized endeavor.[2]

Farmers' adoption of new farm equipment slowed during the period from the Great Depression through the Second World War. Economist Willard Cochrane stated, "These new technologies were not adopted by farmers in the 1930s because their adoption required an outlay of funds which farmers did not have and could not borrow."[3] J. L. Anderson notes that this trend persisted during the Second World War, stressing that farmers had few opportunities to invest in new equipment.[4] Because of economic setbacks followed by war, horse farming persisted in the Corn Belt well into the mid-1940s. Agricultural historian Gilbert Fite eloquently expressed the slower pace of farming at the time: "The three-mile an hour gait of the horse established the speed and power of most field work on farms."[5] Lawrence Schmitz summarized his childhood memories of horse farming by saying, "You wondered how they got any farming done," since horses required feeding, resting, and time to harness and unharness them.[6]

Another wave of technological change immediately followed the Second World War as economic circumstances improved. Farmers in the nation's heartland purchased tractors, new harvesting equipment, grain-handling devices, and hay machinery and adopted new storage practices at a dramatic rate. Between 1940 and 1950, the number of tractors on farms rose from 1.6 million to 3.4 million, signaling the decline of the horse and mule as a power source.[7] By the 1950s, the oddity of using horses on the farm had become newsworthy. In 1956, O. C. Hulett, of the *Racine Journal Times* commented on Tom Ballack, a Wisconsin farmer who still depended on horses to plow and cultivate his fields, stating, "Tom is perhaps the only farmer in Racine County who uses horses regularly for farmwork."[8]

Farmers' abandonment of the horse and the transition to tractor farming was but one of a multitude of shifts to mechanical methods. Farmers acquired new harvesting equipment as well, purchasing mechanical corn pickers, for example, at an especially rapid rate. In the decade between 1941 and 1951, the number of corn pickers farmers were using grew from approximately 120,000 to 502,000. The majority of these machines could be found in midwestern states such as Iowa, Illinois, and Indiana. By 1951, Iowa farmers led the nation in their use of corn pickers, having procured 95,000 machines, followed by their Indiana and Illinois counterparts, who had purchased 48,000 and 80,000 machines, respectively. The device relieved farmers of the physically demanding task of hand harvesting. Charles Scranton, an agricultural engineer, celebrated their introduction: "These pickers have ended the backbreaking and time consuming job of hand harvesting, and the day of the horse and wagon in corn picking is pretty well on the way out."[9] Nevertheless, just as farmers had almost universally adopted the corn picker, the combine, which streamlined the harvesting process by simultaneously harvesting and shelling the corn, was beginning to transform the corn harvest.

Farmers largely abandoned the "old binding and threshing tools" and adopted the combine. By the close of the 1950s, Iowa farmers alone possessed 90,027 of these machines.[10] This was soon followed by the introduction of the combine corn head, the front part of the machine that initiates the harvesting process, which earned a similarly rapid acceptance and further increased the efficiency and ease of the corn harvest.[11] The combine increased both the safety and the speed of processing the harvest since these machines clogged less and gathered corn more quickly than did corn pickers. In the mid-1960s, an *Implement and Tractor* contributor praised the continued improvement of the combine, emphasizing their greater power, ease of operation, and speed in harvesting crops as advantages over earlier harvesting equipment.[12] J. L. Anderson summarized the transformative nature of the combine's impact on harvesting crops by stating, "With the development of the self-propelled combine designed for the corn crop as well as small grains and soybeans, it was possible to own one machine to harvest all the seed crops grown on a farm."[13] Farmers at the time were not only able to harvest crops more quickly, but handling and transport of commodities too became dramatically less burdensome. For instance, the numbers of elevators had increased rapidly between 1940 and 1959, having "doubled on the average every two years," reaching one million on the nation's farms by 1959.[14] By the end of the 1960s, approximately one and a half million elevators on the nation's farmsteads testified to the farmers' acceptance of labor-saving devices.[15] Farmers also benefited from the introduction of many more types of wagons than were available to their early twentieth-century forerunners, including ensilage, side dump, and stack wagons. The wagons of the postwar era also possessed hydraulic lifts and augers, which greatly added to their usefulness and flexibility. In 1978, a National Safety Council (NSC) author summarized such dramatic developments and stated that "farm wagons have come a long way from the simple carts and high-wheeled rack and box conveyance of 50 years ago. No longer are wagons just boxes or flat beds mounted on wheels. They come in many shapes and sizes, often with specialized features, making them suitable for numerous farm hauling and materials-handling jobs."[16] Such

Farmers benefited from labor-saving devices, such as elevators and augers, which greatly eased the transportation of grain, hay, and livestock feed. Such machines also had the potential to cause injury, however, either from their rapidly moving and sometimes insufficiently guarded parts or because they sometimes fell, crushing operators or bystanders. *A.M. "Pete" Wettach Collection, State Historical Society of Iowa, Iowa City. Reproduced with permission.*

improvements taken together reduced the need for human labor on Corn Belt farms in unprecedented ways.

MAINTENANCE DEMANDS

Bruce L. Gardner, a prominent agricultural economist employed at the University of Maryland, summarized this rapid farm mechanization in the mid-twentieth century by stating that "in the forty-five-year period from 1915 to 1960, the transition from animal to mechanical power was completed." Farm families who lived during this period of sudden change were required to adapt quickly to new types of mechanical devices with little or no assistance from outsiders.[17] Safety advocates and some farmers recognized the variety and significance of the challenges operators faced in attempting to learn to manage new agricultural machines safely. Such obstacles included the intense rate of mechanization, the great number and types of machines, and the lack of formal training opportunities.

Some farmers who lived through this transitional era found the process of adaptation to be demanding, especially older farmers. In 1981, Robert Westpfahl, an Illinois farmer, reflected on the difficult shift his family had experienced in going from horse-powered machinery to tractor-powered implements. Farmers of an earlier era "had been walking behind horses for 40 years and all of a sudden they give him this machine, and he knew that before all he had to [do] was tell the horse to stop and the machine stopped whatever he was working. Well, after he got the tractor, he would stop the tractor, but sometimes the machine wouldn't stop. Well, that is just plain suicide because those guys have never been around a machine and mechanical things."[18] Lawrence Schmitz made similar comments regarding his observations of a hired hand who awkwardly made the shift from horses to tractors in the 1940s. Schmitz stated years later, "He'd get on that tractor, and he'd never driven one before, so it was comical to watch him try to get going. It was almost like you had to get down there and lead him."[19]

Some safety writers echoed such perspectives, believing that the mechanical transformation had outpaced farmers' capacity to adapt to the new machines and that farmers needed to pay greater attention to safety. In 1953, Robert Howey, a vocational agriculture instructor at Sycamore High School, in Sycamore, Illinois, indicated that agriculture's rapid mechanization demanded that farmers be trained to run their machinery safely. He stressed that farmers had enthusiastically embraced new technologies, but they had been less excited about adopting proper precautionary practices. He went on, "They are working with machines which are new to them as well as [to] their sons. They do not have the knowledge to teach their sons rules of safety that apply to this machine age."[20] Farmers were less uniform in their assessment of the difficulties associated with technological adaptation, however. James Wilson, for example, compared his adjustment to new farm equipment to the evolution of automobiles. He emphasized that once he learned the basic controls, it was not difficult to incorporate the new equipment into his operation.[21] Such diverging opinions reveal that farmers and those in the vocational agricultural field did not always agree in their views on the ease of adapting to the disruptive nature of new machinery.

The speed of technological change was matched by the diversity in types of equipment that were altering people's work on Corn Belt farms. Farmers needed in-depth knowledge to properly operate and maintain these diverse types of machines. They were often not experts on any individual piece of equipment but were instead generalists who operated and maintained the many different kinds of machinery needed to complete their daily chores. Some writers emphasized that farmers existed in more mechanically diverse environments than did workers in many other occupations. In comparison to the repetitive and limited number of skills required of many factory workers, for example, farm families were called upon to learn many kinds of equipment. This meant that they often never developed the level of proficiency working with their many machines that typical factory workers acquired when using fewer devices.[22]

Agricultural writers commented on the lack of training opportunities to address the great demands brought about by this diversity and noted that this amplified safety risks. In 1942, Frank Zink, the director of the Farm Equipment Institute's Research Department, pondered, "Who trains the operator of a machine? Who trains the new or young operator of a farm machine? The probable answers to these two questions are—the older persons, either fathers or farm hands, neither class of which may have had any safety training."[23] In 1968, Jack Burke, an NSC staff member, wrote in an article entitled "The Lonesome Plowman" that, unlike many other workers, farmers were exclusively responsible for their safety. He stressed that airline pilots, cab drivers, and a host of other individuals were forced to undertake formal training prior to flying a plane or driving a car or truck, whereas most farmers enjoyed very little guidance before using their equipment. Burke indicated that most farmers had minimal tractor instruction except for that given by older family members, implement dealers, or another farmer. He wrote of the farmer faced with new equipment, "What help does he have? He needs neither license nor physical exam. He may be almost any age, from a green fledgling eight-year-old straining to reach controls to a doddering, half-blind octogenarian with reaction time approaching that of a drowsy mule."[24] Many farm safety advocates believed that parental guidance in the realm of safety was insufficient in addressing hazards and that farmers needed the help of outside experts to learn to run their operations safety and efficiently. Such opinions contributed to the creation of the farm safety profession during the era.

Farmers not only operated many kinds of equipment, they were also burdened with significant maintenance responsibilities that, if neglected, increased the chance of an accident. Such tasks appear to have become more burdensome as agricultural equipment became increasingly sophisticated. Maynard Coe, director of the National Safety Council's farm division, recognized these issues, stating that their "mechanic" role alone might involve welding, electrical wiring, and even proper lightning rod installation.[25] In September of 1970, Charlie Nettles, in an article entitled "Farmers Tell Machinery Gripes" appearing in the *Des Moines Register,* described farmers' frustrations with machinery, particularly when it malfunctioned during planting and harvesting. Nettles also revealed that farmers frequently made equipment "modifications," such as adding gas tanks and balancing weights to combines. These alterations illustrate the independence farmers exerted in their workplace and how much their safety depended upon their own mechanical proficiency.[26]

A farmer's maintenance role was particularly crucial in regard to proper machinery upkeep. Farmers were advised to read operation instructions since installing the wrong parts for a machine could be dangerous. In 1943, a writer for *Hoard's Dairyman* cautioned farmers about being too zealous in their improvisation, stressing that only approved parts should be used for equipment and that hasty "baling wire repairs" could result in dangerous mishaps. In 1952, Charles Scranton warned farmers against operating their corn pickers without thoroughly consulting their equipment manuals, noting that the booklet contained a wealth of information essential for proper operation.[27] Farmers often had to stay abreast of changing maintenance requirements for their equipment to run properly. Pat Larkin remembered a time when "you would stop in the middle of the field and take the carburetor off the tractor, clean it out, start it up and it would run pretty good. Now you have to have a special mechanic."[28] The trend toward greater mechanical complexity seems to have only intensified as the century wore on. Larkin's perspective shows how twentieth-century farmers' maintenance needs and safety were becoming increasingly dependent on operating manual writers, equipment designers, and mechanics. His insights show that although farmers perceived themselves as being independent, their lives and work were deeply affected by the decisions of agricultural engineers and the implement companies that employed them.

Thus, farmers who lacked proper knowledge about or neglected upkeep of their equipment could potentially jeopardize their own safety or the safety of their family members. Safety advocates cautioned farmers to check their tractor brakes regularly, for example, because brake failure could result in accidents. Even comparatively simple machines such as wagons required regular maintenance, since hauling equipment evolved into complicated machinery as the decades progressed. Farmers needed to check wagons to make sure they could withstand heavy loads, to guard moving parts with shielding, and to perform regular lubrication to ensure proper functioning as well as overall safety.[29] Perry Middlesworth remembered operating a John Deere 60 tractor that had a broken hand clutch. He "freewheeled backwards down a hill and the brakes didn't work very well."[30] He said his dad promptly repaired both the clutch and the brakes out of concern for his son's safety. Such situations were most likely not unusual since some farmers sometimes used older equipment that needed frequent upkeep. Farmers' safety was clearly deeply affected by their unique financial situations.

A farmer could prevent accidents and experience a more efficient harvest if he properly looked after his corn picker or combine. In 1955, Daniel Kitchen, an NSC agricultural engineer stressed the importance of safety practices, stating that "[a] well-adjusted machine in good repair gives less trouble [and] is less likely to tempt the operator to take chances."[31] Proper use of a corn picker required repeatedly checking for damaged components, making adjustments, and avoiding excessive speeds. Other necessary procedures included greasing the machine's parts and making clutch modifications. Safety advocates recommended that machinery be maintained in these ways even after the harvest so the farm equipment would be ready for next season.[32]

Farmers also needed to be attentive to variables such as harsh field conditions when operating corn pickers and combines. Charles J. Scranton stressed that, even in a single field, corn plants could present great diversity due to variations in moisture and soil. Plant height and density could vary greatly, he noted; some corn plants stood over twelve feet high and others not much more than half that size, while one field could have both sparse and thick patches. A weedy field could also affect how easily the corn picker would clog, making it necessary for the farmer to adjust the picker to compensate. Safety writers also pointed to field conditions as a factor in preventing combine-related accidents. In the early 1970s, Joseph F. Gering, an Oklahoma Extension safety specialist, stressed that farmers should practice adequate weed management measures since such noxious plants became entangled into a combine's many moving parts and impeded smooth operation. He also suggested that prior to harvesting farmers should inspect their fields for stones and other debris that could damage equipment.[33]

Clearly farmers worked under conditions that lacked the uniformity in materials and tools used and in procedures possible in factories. Observers and farmers accepted that the distinctiveness of the agricultural setting precluded factory-like efficiency, giving additional impetus to the movement for farm safety education.

AN ABUNDANCE OF HAZARDS

The mechanization of Corn Belt agriculture brought with it the possibility of death and serious injury in a myriad of forms. The most common causes of injury or death while using farm equipment, however, were being crushed by the machinery's overwhelming weight or being pulled into rapidly moving machinery parts. A farmer could be trapped under a tractor as a result of a rollover; a child might be caught in a power take-off or fall and be crushed underneath a tractor's wheel. Corn-picker operators needed to be aware of the many fast-moving components that could sever a finger or crush an arm. Processing hay also presented hazards: An operator might get caught in a baler's fast-moving components or might tip over a tractor while moving the large round bales of hay, suffering serious injury or death. Farmers' use of new material handling and storage devices for moving and storing grain, such as elevators and augers, posed dangers due to the machines' many quickly moving mechanical elements, but also because the machines could collapse, crushing an adult or child beneath them. Sadly, these examples do not represent an all-inclusive list of the hazards farmers faced, but they do serve to show that Corn

Belt farming in the mid-twentieth century was a potentially lethal endeavor.[34]

An Iowa Cooperative Extension Service study covering the period from 1947 to 1971 helps place new technologies and farm injuries in a broader perspective. The writer concluded that "the rapid rise in the use of machinery on farms has increased the exposure of farm persons to hazards of machine operation. The problem of exposure compounded because the total farm population was decreasing while the number of machines per farm was increasing. The result is that fewer persons were operating more equipment." The author revealed the machinery farm fatality toll, stating that "a total of 1,773 Iowans have died as a result of accidents with tractors and farm-related machinery." The researcher also indicated that "over two-thirds of the total accidents have taken place on the farm, with an additional 25 percent occurring on highways and roads."[35] Lawrence Schmitz recalled this pervasive nature of farm dangers indicating that his father "warned him daily" about the many hazards associated with the agricultural lifestyle.[36]

THE DANGERS OF LARGE, HEAVY EQUIPMENT

The industrialization of agriculture affected both the size and the types of equipment that farmers used. Farm machinery reflected the increasing scale of agriculture as it grew in size and weight as farmers cultivated more land with fewer workers. Farm owners often compensated for the escalating costs of operating their farms by acquiring more land and purchasing larger, more sophisticated equipment.[37] Therefore, machinery such as tractors, elevators, wagons, and other devices grew ever more massive as the twentieth century wore on. People who found themselves under such equipment as a result of an overturned tractor or a collapsing elevator faced serious injury or even death.

Clearly, although tractors undoubtedly contributed to rising levels of productivity on mid-twentieth-century farms, they could sometimes be the cause of farming accidents.[38] The majority of farmers of course avoided death while operating their tractors, but close calls were common. Larry Glenn recalled that he almost overturned his tractor when he accidentally ran it into a ditch. Glenn admitted that he probably wasn't paying close enough attention, but fortunately he was unscathed.[39] Jim Klein also had a narrow escape while operating a tractor because he wasn't focused on driving: "So much of the work is kind of mundane, and it is easy to switch off your brain or get distracted, but just about the time you do that, that's usually when something goes wrong."[40] Such comments illustrate how a lapse of judgment or an unusual circumstance could lead to an overturned tractor and the risk of being crushed to death or seriously injured by thousands of pounds of steel.

Tractor operators needed to be aware of a variety of potentially dangerous situations, such as excessive speeds, going up severe slopes, or pulling heavy loads downhill, all of which could result in close calls, severe injury, or even death in turnovers. Tractor capsizes were particularly dangerous prior to the widespread use of rollover protective structures (ROPS) that now shelter drivers. In April 1952, Paul Murphy, a Marion County, Iowa, farmer, experienced a frightful incident. His tractor overturned on a hill while he was working his fields, but fortunately he was able to quickly leap off his tractor and escape without injury.[41] Rex Peterson remembered that once his dad "turned a tractor over on the backside of a terrace," but fortunately he too escaped serious injury.[42] Unluckily, tractor rollovers weren't the only way one could be crushed or injured by a tractor's weight. Don Rimathe recalled that he knew a boy "who fell out of the loader bucket and got run over by a Farmall H, but he's still here to talk about it."[43]

Rural families also faced falling or crushing injuries while using their tractors to pull wagons filled with grain, hay, and other materials. In 1968, an NSC safety publication entitled *How to Get Your Wagon Rolling Safely* indicated that although wagon-related deaths were rare, injuries were fairly common. Farm accident studies conducted in Michigan and Ohio indicated that wagons fell just below tractors and elevators as leading accident agents. Although individuals could be harmed in such mishaps in a number of ways, falls appeared to be a leading cause of injury. The NSC authors indicated that "falls accounted for about 60 percent of wagon injuries. The remainder resulted from being run over or being caught between the wagon and a building or tool."[44] In the late 1970s, NSC farm accident researchers confirmed this view of the prevalence of wagon-related falls through more comprehensive farm accident studies.[45]

These researchers indicated the variety of reasons a person might tumble from a wagon, including "standing on its tongue, unwisely riding on a wagon, or losing one's footing on grain or other slippery materials. People who carelessly rode atop wagons were especially susceptible to falls. The likelihood of such an accident increased if the tractor operator started unexpectedly, stopped suddenly, or behaved in other unpredictable ways." The NSC researchers further stated, "Many injuries happen to wagon riders during loading operations because they fall on or off the wagon." Children onboard as second riders were sometimes seriously hurt or even killed if they fell off a tractor and were crushed by wagon wheels.[46] Farmers also had to be aware of the dangers of overturned wagons, a situation encountered by Jim Klein's brother. Klein, along with his dad and brother, tipped a wagon while in the process of unloading ear corn on uneven ground. He stated that his brother "walked right beside it and it missed him by six inches. Nobody saw it coming. If he would have been slower or farther behind, he would have been underneath it."[47] Klein's emphasis on the ground's unevenness again highlights the important point that farm safety was influenced by farming's outdoor environment.

The process of haymaking was also affected by the increasing scale of agriculture as new labor-saving devices reduced physical labor but created new safety hazards. One of the more revolutionary changes in haymaking was the advent of the "big round baler." By the mid-1970s, such devices produced hay bales as large as 1,000 to 1,500 pounds. Sam Brungardt, an *Implement and Tractor* contributor, stressed that farmers were sometimes unwisely adapting older front-end loaders for moving these large round bales. He stressed that the considerable weight of the bales could lead to great instability during transport and contribute to tractor overturns.[48] Although safe, large bale-handling equipment was rapidly entering the marketplace, farmers did not necessarily immediately purchase the safer transportation devices. Farm innovations did not proceed uniformly, and the willingness of agriculturalists to modify old equipment for new uses often multiplied risks.[49]

Youthful inexperience or lightheartedness could also sometimes pose crush dangers while using heavy machinery such as tractors and elevators. Lawrence Schmitz related a story that occurred in the late 1940s. He and his friend were headed home on their tractors, "out of sight of both dads," when they kicked their tractors out of gear and raced down a big hill. He stressed that such races were unsafe and

Farm family members were sometimes injured in falls. Large wagon loads, uneven ground, and unstable materials could be contributing factors. In addition, tractor drivers usually did not have rollover protection throughout much of the mid-twentieth century. *A.M. "Pete" Wettach Collection, State Historical Society of Iowa, Iowa City. Reproduced with permission.*

could have resulted in an overturn, but he also noted that such potentially dangerous playfulness was not unusual for farm boys.[50] Perry Middlesworth remembered how once as a boy he had thought he was "strong enough to lift up an elevator. It came down" on him and pinned him to the ground, leaving a gash in his knee.[51] Such examples reinforce the point that age was no barrier against exposure to mechanical hazards. Young people especially sometimes took unnecessary risks or were inadequately trained or prepared to operate heavy equipment that could crush them and possibly end their lives.

FAST-MOVING MACHINERY

The industrialization of agriculture meant that farmers were rapidly adopting many types of equipment possessing fast-moving parts. The power take-off, corn picker, combine, elevator, and auger were only some of the mechanical innovations and technologies that posed particularly great danger. Operators caught in the many moving parts of one of these devices could lose a limb or their lives. The tractor's power take-off (PTO), a spinning shaft which passes energy from a tractor's engine to a variety of equipment, could be especially hazardous. A person could become entangled in a PTO by accidentally falling on the whirling shaft, by attempting to step over it too hastily, or by wearing loose clothing that becomes caught in it. All of these unwise practices could result in serious injury or even death. Removing the PTO guard multiplied the hazards. In 1950, E. W. Tanquary, chairman of the Farm Equipment Institute's Advisory Engineering Committee, stressed how important it was for farmers to use such shields, stating that "[t]he best designed and most expensive shield ever provided is worthless if left in the implement shed and the implement operated without it, yet accidents are reported where ample shields were provided with the implement and left off through carelessness."[52]

Unfortunately, many machines, such as elevators and augers, lacked proper guarding throughout the period. In 1969, David McFarland and William Fletcher confirmed this in a 1969 investigation of augers and elevators currently being sold in the United States. They indicated that many machines either lacked guarding in critical areas or had insufficient guarding for the operator's protection.[53] In 1972, Norval Wardle confirmed such observations after reviewing a number of manufacturing brochures; he surmised that such machines were on the whole insufficiently guarded. Wardle further illustrated his point by disclosing that some farmers had crafted guards themselves following accidents. A particularly powerful example involved a Kentucky farmer who added shields to his auger after a girl had lost a foot in the machine. Another situation involved Horace Neu, a farmer from Nevada, Iowa, who, after catching his hand in an auger, fashioned a "lattice type guard over the auger's intake."[54] Wesley Longnecker reported that his son got his foot "caught in an auger and lost a toe." He believed that insufficient shielding contributed to the accident. Such incidents reveal how quickly one could become seriously injured in such machines if they were inadequately guarded.[55]

Rex Peterson was also keenly conscious of the hazards associated with moving parts and recalled the time his son had stepped over a PTO shaft and it "scared him to death." Peterson said that fear coursed through his body so completely that he broke the handle of a corn rake he was holding as he watched his son's unwise action.[56] Perry

Middlesworth remembered being extra careful around the PTO shaft as a result of seeing a sack become entangled in the whirling shaft while using a grinder-mixer. He stated, "It caught a burlap bag once, and in no time it was wrapped around it tight." He went on to say that after seeing that bag he was "a lot more safety oriented."[57] Farmers' seasonal challenges were also influenced by PTO hazards. In 1966, L. N. Knapp of the Institute of Agricultural Medicine indicated that such hazards increased in cold weather when farmers donned winter clothing that could more easily get caught in a whirling PTO shaft.[58] Knapp's comment affirms that the outdoor work environment made the industrialization of farming distinct from other industrialized settings.

Corn pickers too sometimes caught or pulled a farmer to death or serious injury. Farmers were particularly vulnerable when attempting to unclog a picker or to make repairs with the power on while rushing to complete the harvest. In 1952, the wife of a Wapello County, Iowa, farmer commented on this problem, stating that "I think that the men just try to get in a hurry when they are tired."[59] An exhausted farmer could get caught in the picker's rollers or chains in a variety of ways. In 1955, Daniel Kitchen stated, "Many accident victims have dismounted with the power take-off running with no intention of approaching the picker. Then they saw an ear crossway in the elevator, saw a stalk of corn laying across the picker points—or they stumbled or lost balance. And another hand was added to the list." Regardless of the situation, Kitchen reinforced the idea that farmers could prevent many accidents by turning off machines prior to attempting to fix them.[60] Despite such advice, harvesting accidents exacted a grim toll. Leroy Hogeland indicated that he often purchased used equipment at farm sales. He stated that sometimes "you'd go to a farm sale and you'd see a guy with just one hand and you didn't have to ask him what had happened; you knew."[61]

Farmers and their families who experienced such accidents carried the memory for the rest of their lives. Rex Peterson reflected upon the many ways his life changed after his dad lost his arm in a corn picker in the late 1940s. Peterson's dad had been applying oil to the corn-picker's drive chain with the machine running idle when it "caught his coat, twisting it so tight that although it didn't break any bones, it shut off his circulation," and he subsequently lost his arm.[62] Despite such accidents, harvesting equipment remained potentially dangerous well into the 1970s. Bernard Stodghill vividly recalled his corn-picker accident forty-five years later, demonstrating the life-changing dimensions of such accidents. He remembered how he caught his hand in the corn-picker's steel roller and lost most of a finger while harvesting corn late at night.[63] James Wilson provided a chilling account of a near-fatal experience with a baler in 1967. He stated, "I always felt like I was 4 or 5 seconds away from losing my life. I got my foot caught in a [small round] Allis Chalmers baler and the tractor was running wide open." He surmised that if a hired man hadn't been present, the outcome might have been very different.[64]

Such near-death experiences reveal that the tremendous benefits brought about by such machines were not without a cost. The introduction of new machines contributed to the creation of a more technologically complex work environment. Therefore, farmers faced a situation that required increasingly specialized and technical knowledge to run their operations efficiently and safely. This meant that farmers were deeply affected by the design decisions of agricultural engineers and farm implement company employees who never set foot on their farms. In addition, diverse technologies such as tractors, wagons, elevators, and a host of other devices presented very real dangers to their operators. Such hazards were amplified by many factors, including the diverse working environments faced by farm families, the families' relative independence, and the rapid rate of technological change. The historical evidence also suggests that demands of time and money as well as the multifaceted nature of the work complicated farmers' efforts to work safely. Farmers facing considerable financial strain were sometimes tempted to place a greater priority on getting their work done promptly than on working safely. The lack of safety features, such as adequate guarding and tractor roll bars, amplified the dangers. Although the miracle of increasing levels of agricultural productivity owed much to mechanical improvements, farmers' mishaps with these machines also resulted in a harvest of human misery, a harvest that has largely been ignored both by the urban public and by historians.

4

THE PERILS OF CHEMICAL USE

FOLLOWING THE SECOND WORLD WAR, farmers not only had to adapt to mechanical change, they also embraced an ever-expanding repertoire of chemicals in their operations. Corn Belt farmers immediately recognized the benefits of employing chemical remedies against insect, weed, and soil challenges. James Wilson remembered, "We were tickled to death with the chemicals because that meant that we didn't have to walk that bean field" to pull all the weeds.[1] Leroy Hogeland concurred, "The first time I used Atrazine, I thought it was the greatest thing since sliced bread especially since it worked well against Giant Foxtail."[2] John Van Ringelestein remembered an occasion on which a landowner from whom he rented first recognized the power of applying commercial fertilizer. Van Ringelestein recalled that although the landowner initially didn't want to pay for it, once he saw the dramatic increase in yields, "he wanted to put fertilizer on everything."[3] These comments suggest that farmers accepted agricultural chemicals as they did new mechanical improvements. Both were viewed as necessary tools to increase productivity, increase efficiency, and reduce the age-old drudgery of more organic forms of farming.[4]

Industrialization of agriculture not only involved the work of mechanical engineers but also the seemingly miraculous substances of the chemist's lab. Agricultural chemical researchers promoted their creations as powerful tools to suppress pests and weeds in the ongoing struggle to increase productivity. Many farmers thought that DDT represented the universal remedy in the ancient battle against crop-consuming insects. Anhydrous ammonia and other fertilizer innovations were widely viewed as welcome tools for increasing crop yields.[5] Farmers also enjoyed the weed-killing benefits of 2, 4-D, which essentially caused weeds "to grow themselves to death." Unfortunately, such innovations destroyed their intended targets but also possessed the potential to harm others. Midwestern residents would learn that these chemicals had the potential to harm consumers, the environment, and the farmers who applied them.

Although farmers had used pesticides such as arsenic and cyanide for decades in their fight against insect pests, chemical options expanded greatly following World War II. In 1959, J. M. Magner, an entomologist employed by the Monsanto Corporation, stated that although pesticides had been available since the nineteenth century, the emergence of "organic pesticides" during the 1940s greatly multiplied such weaponry. In 1961, Mary K. Farinholt, an agricultural chemical specialist, provided a numerical understanding of this transformation in pest control. She stated that in that year alone there were "approximately 94,000 brand-registered pesticides on the market" and that American growers had purchased around "500 million pounds of pesticides worth $300 million."[6]

Many agricultural writers believed pesticides were indispensable in eliminating the threat that pests posed to meeting the expanding global population's food needs. In 1965, a *Farm Safety Review* writer indicated the problem's scope, writing that "ten thousand kinds of insects" threatened crops and livestock in the United States.[7] In 1967, a USDA employee laid out the likelihood that farmers would come to rely long-term on chemicals, stating his belief that no non-chemical remedy was effective against the pests that, despite the aid of such substances, still consumed almost "a third of our potential production." Magner also justified growing chemical use to meet the challenges of an exploding world population and to compensate for the continued loss of farmland to urban expansion.[8] Perhaps the strongest evidence regarding the widespread acceptance of chemical control methods was that in 1962 only 2 percent of the nation's entomologists were researching alternative insect control methods.[9]

Farmers in the Corn Belt also participated in unprecedented changes in fertilizer practices, especially with the introduction of anhydrous ammonia.[10] In 1955, Iowa farmers deposited 18,000 tons of the fertilizer on their fields, a figure that rose to 348,000 tons just twelve years later.[11] Farmers' increasing use of anhydrous ammonia was also affected by the ever-lowering cost of producing the fertilizer. Bruce Gardner noted that the price of the fertilizer fell "by about one-half in nominal terms between 1960 and 1970."[12] More recently, J. L. Anderson described anhydrous ammonia as one of the least expensive and most potent fertilizers, factors contributing to its widespread use after the Second World War.[13] Industrialization of Corn Belt agriculture during the period was crossing many technological realms that, taken together, profoundly reconfigured farmwork.

Agricultural chemical companies not only produced an impressive variety of chemicals, they also developed clever advertising campaigns to entice farmers to use their products. The nation's extension agents and other government officials also encouraged growers to increase their use of substances as the preferred method of controlling insects and weeds.[14] However, Lawrence Schmitz implied that he and other farmers were not made fully aware of the hazards associated with agricultural chemicals. In reference to exposure to several chemicals, such as Atrazine and Treflan, he stated, "They promoted these chemicals as perfectly safe products." He elaborated that although he is now aware of hazards associated with these chemicals, he wasn't so informed during the 1960s.[15] Bernard Stodghill also believes that an emphasis on chemical safety was lacking in the early years and that most farmers were more concerned with application rates and procedures than with safety considerations.[16] These labor-saving technologies were clearly welcomed into use, however, even though users did not fully understand their properties and potential unintended consequences.

CHEMICAL HAZARDS

Improper use of chemicals could have serious consequences, including injuring the user or family members and neighbors or poisoning the nation's food supply. Despite the possibility that people could experience harm from these chemicals in a variety of ways, the most common avenues for poisoning were breathing in the substances, being exposed through the skin, and swallowing the chemicals.[17] Safety experts believed that farmers lacked proper training and knowledge in the use of these powerful new technologies, and they promoted the idea of specialized education in how to use them correctly. The situation was made more dangerous because federal and state regulations governing the registration and use of agricultural chemicals were rather lenient during the era, contributing to the proliferation of chemical hazards throughout the Midwest.

The growing number of chemicals used in cultivating crops posed a significant number of health risks. If farmers incorrectly used herbicides, they could inadvertently harm their harvests. Researchers also eventually discovered that pesticides such as DDT could accumulate in living organisms and consequently pose hazards to both animals and people.[18] When farmers accidentally released anhydrous ammonia, it could result in serious injury, such as blindness or even death. Farmers who did not wear proper protective equipment or carefully read product labels could suffer serious illness or injury. For instance, DDT caused considerable health problems, as evidenced by the 338 people in the United States documented as being poisoned by it in 1951 alone.[19]

Although available statistics reveal a serious problem, the fragmentary numerical record does not provide a complete picture. Researchers repeatedly emphasized the challenge in determining the number of chemical-related farm accidents. In 1962, Farinholt stressed that data collection regarding injuries and deaths associated with pesticides lacked standardization. Nevertheless, she indicated that specialists had calculated from existing data that 166 Americans had been killed in pesticide accidents in 1959. Farinholt emphasized that nonlethal incidents might be as much as one hundred times as common as deadly incidents.[20] During the same year, Keith Long, the head toxicologist at the University of Iowa's Institute of Agricultural Medicine, echoed Farinholt's statements regarding insufficient research.[21] Such observations were both widespread and persistent throughout the era.

In 1972, William Fletcher, a National Safety Council (NSC) agricultural safety engineer, stressed that those wishing to understand the agricultural chemical safety problem still lacked sufficient data. He stated that part of the difficulty lay with researchers' local, not national investigations. He revealed that only California health officials had kept accurate chemical accident records, showing a total of 163 fatalities related to agricultural chemical accidents between 1951 and 1969.[22] The state's annual toll of illnesses caused by agricultural chemicals was much higher. In 1971, "1,284 reported cases" of illness were attributed to such substances.[23] Even as late as 1976, the author of an NSC rural accident prevention bulletin decried the lack both of data and of literature on the subject.[24]

Nevertheless, many safety experts held high expectations for farmers' safety knowledge. A *Farm Safety Review* writer emphasized that although all chemicals had to pass a governmental approval process, once purchased, the farmer became fully responsible for using them safely.[25] In 1955, Harold Heldreth stated, "The farmer should fully understand and be able to explain to his workers the dangers involved, protective clothing necessary, antidote, first aid treatment and proper methods of handling pesticides."[26] Unfortunately, the time and money required to follow such prudent advice was considerable, and in some cases impractical. This was especially true in the Corn Belt, where farm families were agricultural generalists with limited time and energy. Thus, for most of the mid-twentieth century, farmers were given the weighty burden of providing for their own chemical safety, even though few possessed formal chemical safety training. In addition, farmers' experiences with chemicals also differed greatly from those of workers in other occupational settings. In 1957, Henry Doyle, an official for the U.S. Public Health Service, argued that unlike chemical use in manufacturing, agricultural chemical use was much more diverse and less controlled.[27]

Such large, corporate commitments were unrealistic for family farmers in the Corn Belt, who lacked both the financial resources and the expertise possessed by large chemical factory owners. The farmers' rural setting also complicated both preventative and emergency measures. Dr. Clyde Berry, associate director of the University of Iowa's Institute of Agricultural Medicine, while speaking at the Society of Agricultural Engineers national conference, suggested that many farmers did not have access to necessary support services. He stressed that some farmers lived in communities where respirators were not sold by area businesses and even access to water might be limited in the isolated fields where accidents could occur; long distances to hospitals and clinics delayed emergency medical treatment.[28]

Properly pouring and mixing chemicals in the mid-twentieth century posed a complex of problems similar to those in other areas of rapid farm industrialization. In 1969, William Fletcher recommended that farmers maintain an especially high level of awareness when mixing chemicals, a process during which they were particularly vulnerable to accidental contact. Correct mixing practices were also required for equipment to function properly. Fletcher stressed the importance of using the proper sequence when dispensing each substance into equipment tanks, since incorrect mixing "can result in the clogging of nozzles, frequent clean-out and extra exposure of the operator to chemicals."[29] Such advice was sound, but like many accident prevention measures, it took time and knowledge to execute, neither of which was always easy for farmers to obtain.

Some farmers were wary of pouring and mixing chemicals. Bernard Stodghill emphasized this point when recalling his use of Treflan in the 1960s. Stodghill revealed that he "didn't use rubber gloves when he mixed it" and that the small capacity of his spraying system's tank meant that he had to frequently blend the product. These numerous mixings meant he was "handling it all the time," and he knew that the "odds were against you" for avoiding chemical exposure.[30] Perry Middlesworth also realized that the task of mixing chemicals could expose farmers to hazards. He recalled that once his father-in-law was dispensing a chemical into a tank and "it splashed onto his face."[31] Such examples strengthen Fletcher's point that farmers were especially vulnerable while blending and pouring agricultural chemicals. These observations confirm that farms often lacked the uniformity in equipment and safety practices such as those found in manufacturing settings.

Although farmers' contact with chemicals from splashing appears to have been most common in the mixing process, some farmers were heavily exposed while applying chemicals. Pat Larkin stated, "I used a sprayer and the barrel sat right behind me and that thing splashed on my back until it was half empty."[32] James Wilson remembered a dramatic incident with a crop duster in the early 1960s. He described how he had several fields that had been overwhelmed by weeds due to an especially rainy spring. He employed a pilot to treat the fields, and the pilot told him to stand out in the field and "flag for him, but the wind happened to be in the wrong direction and when he was done, I was just soaked with moisture."[33] Jim Klein recalled that his family didn't use protective gear such as rubber gloves while applying chemicals and often had lunch or snacks without washing their hands.[34] Such examples dramatically illustrate the various hazards associated with working with agricultural chemicals.

Farm chemical dangers were also deeply influenced by the familial nature of Corn Belt agriculture, as evidenced by the youths who died or were injured in chemical mishaps. A number of studies and examples confirm this conclusion. The authors of a Missouri Extension pamphlet indicated that 415 out of 569 poisoning victims treated at the state's poison control centers in 1975 were youths five years old or younger.[35] Children were almost always killed or injured by chemicals due to adult inattention. This was particularly true regarding improper storage practices.[36] One tragic case involved a six-year-old child. The boy discovered a discarded jug of the insecticide tetraethyl pyrophosphate (TEPP), an early and potent pesticide, while playing in an orchard and accidentally spilled it onto his legs. He struggled home but although his father promptly took him to the hospital, he died six days later. Ruth Falck also recalled how chemical use and children's play intersected on their Iowa farm. Her daughter sometimes played with chemical bottle caps used in their dairy operation: "We made sure they were rinsed, but we probably would have been more careful knowing what we know now."[37]

Children's vulnerability to chemical accidents persisted throughout the twentieth century. A 1983 *Wallaces Farmer* article revealed that youngsters faced particular dangers from granular pesticides. Agricultural chemicals in such highly concentrated forms could be easily swallowed. The writer elaborated, "The lethal dose of these granules, if swallowed by a 20 to 30 lb child is in the range of 1/8 to 1/4 of a level teaspoon." Donald Morgan, a physician employed by the University of Iowa's Department of Preventative Medicine, vividly explained the potential lethality of such pellets, noting that a deadly amount could easily cling to a child's sweaty hands. Such poisonings often occurred when the granules were haphazardly spilled on the ground where children could easily find and ingest them. Dr. Morgan illustrated his point by describing one poisoning incident that led to the deaths of two sibling toddlers: The children had been playing outdoors on their Iowa farm and ingested a very small amount of spilled pellets. Morgan concluded his discussion with a number of ways the problem could be avoided, including keeping children away from areas that might contain chemical residues.[38] Such advice would be challenging to follow on Corn Belt farms where the workplace and the home were one and the same.

ANHYDROUS AMMONIA

Although farm families were not fully prepared to address all the chemical safety issues they faced, the technology available for using these new substances became ever more practical. For instance, improvements in delivery and application equipment contributed to farmers' growing use of anhydrous ammonia. Prior to the Second World War, farmers on the West Coast were already using substantial amounts of it by injecting it directly into the water used to irrigate their fields. In the late 1940s, researchers at Mississippi State College had developed application systems that could inject anhydrous ammonia into the soil itself. The process began at factories that produced and stored anhydrous ammonia in large tanks under high pressure. Semitrucks or rail cars transported the fertilizer from its point of origin to distribution centers where farmers purchased it. Fertilizer companies then delivered the anhydrous ammonia fertilizer in nurse tanks; tubes then fed the substance into applicators, which used "sabre-shaped blades" to inject the nitrogen into the earth. Tractors pulled this equipment while shooting the material "at a depth of four to six inches" into the soil. Such ease of use increased the feasibility that farmers could use this particular type of liquid fertilizer.[39]

Nevertheless, farmers' escalating use of the chemical contributed to new risks. The fertilizer's characteristics, which included its high storage pressure, subzero temperature, and toxicity, made it a potentially dangerous material. In 1972, a writer for an OSHA brochure stressed that mishaps often occurred during the transfer of the fertilizer between storage devices. If a person improperly released and shut the valves, the article continued, serious injury to that person and to coworkers might result. Moving a nurse tank without first disengaging the hoses or attempting to fix the "couplers" connecting the tubes could also result in leaks. Such accidental emissions could have disastrous consequences, including severe burns, blindness, or even asphyxiation if inhaled.[40]

The many accident stories that circulated throughout the agricultural community illustrate the potential hazards. Some of these incidents, such as an accident in Crete, Nebraska, in 1969, were spectacular in nature. In this episode, thirty-three thousand gallons of anhydrous ammonia were released into the atmosphere from a wrecked railroad car. The discharge of this tremendous volume of ammonia resulted in six deaths and more than thirty injuries. Although such events were rare, many smaller accidents also occurred. Because such accidents were isolated, even if serious, individual episodes, the prevailing insufficient efforts at collecting accident data most likely masked the true scale of the problem. In 1971, James Hall of Keswick, Iowa, came in contact with the fertilizer as the result of a hose leak that sprayed the ammonia across his face and into his eyes. Arlo Jacobsen, the *Des Moines Register*'s agricultural editor, recounted the young farmer's harrowing experience: "With no emergency water tank to flush his eyes, Hall stumbled half a mile to a neighbor's place for water." Although he recovered his eyesight, the incident left a powerful impression, convincing Hall never to handle the fertilizer again.

Farmers like Hall sometimes escaped serious injury, but they carried their memories of these close calls for years. Don Rimathe recalled one anhydrous accident in which "the draw bar pin, safety chain, and the safety valve" all failed and, as a consequence, a large amount of anhydrous ammonia escaped into the air. He stressed that he was "very lucky," since the wind was blowing away from him and fortunately no one else was around to get hurt. Rimathe commented that although no one was injured, he came close to draining the tank, and it was "a bad situation."[41] Perry Middlesworth also recalled that once, while treating a field, "a hose broke and sprayed. I pushed the tractor in neutral and ran and put my head in the dirt."[42] James Wilson, too, every once in awhile had a small amount leak from a "coupler," noting it would "take your breath away," but fortunately he "never had a hose break."[43] Such observations illustrate the complexity and potential hazards associated with working with this dangerous chemical.

In the early 1970s, a number of studies provided a useful synopsis of the midwestern anhydrous ammonia problem. One especially revealing investigation, conducted by two Purdue University agricultural engineers, J. B. Liljedahl and F. R. Willsey, assisted by Robert C. Rund, an employee of the office of the Indiana State Chemist, looked at anhydrous ammonia accidents in Indiana in 1971 and 1972. Although many incidents were not serious, a substantial number were; among survey respondents, "33 percent classified their injuries as major." Failures of hoses, valves, couplers, and other transmission devices were responsible for many of the leaks, resulting in "81 percent of the accidents." The researchers found, predictably, that every accident occurred in spring and early summer, when farmers applied anhydrous ammonia to their fields.[44] This observation highlights how seasonal pressures affected Corn Belt farmsteads' exposure to hazards.

The Indiana investigations provided additional insights regarding ammonia accidents, including the significant point that most incidents occurred on the farm and not at manufacturing facilities or distributors. Such findings once again demonstrate how ill-prepared farmers were to assume all aspects of responsibility for their own safety in handling a potentially highly dangerous substance. Fletcher argued that the manufacturers and suppliers of anhydrous ammonia appeared to have adequate safety practices but that farmers lacked proper protection when applying the substance to their fields.[45] Researchers also indicated that farmers often transported the ammonia to their own farms, since the investigators found that over half of them "filled the nurse tank at the dealer's place by themselves."[46]

Although farmers generally did not believe they were poorly prepared to handle anhydrous ammonia, the Indiana study confirmed that human error was a major culprit in accidents. Human oversights included improper use of the equipment, neglecting to wear a safety apparatus, or ignoring standard application procedures. The investigation showed that "27 percent of the hose-end valves on the nurse tank were improperly secured during transport" and that individuals involved in accidents rarely used eye protection or gloves.[47] In addition, lack of "training or failure to follow accepted operating procedures were obvious in 27 out of 64 cases studied." Farmers were also untrained in using the complicated features of anhydrous ammonia equipment, further amplifying the dangers.[48] Such findings undoubtedly bolstered safety experts' claims that farmers needed new and specialized knowledge to meet the demands of their increasingly industrialized environment. Such studies further justified the need for farm safety specialists to both research the issue and provide prescriptions to reduce ammonia mishaps.

The fact that safety activists generated a plethora of materials to educate the rural public on the issue reveals their understanding that farmers were inadequately prepared to meet the operation demands of anhydrous ammonia. They advised those who handled the fertilizer to wear safety gear, inspect equipment, use proper operational measures, and understand proper emergency procedures. Goggles and gloves were especially important, since the eyes and hands were particularly susceptible to exposure and injury. In 1972, Arlo Jacobsen advised farmers to carefully and regularly check their application equipment to avoid any potentially dangerous leakages.[49] Many writers also stressed that everyone, including children, should be knowledgeable about how to respond to an anhydrous ammonia emergency. Such advice once more highlights the familial aspects of Corn Belt agriculture.[50]

Experts consistently emphasized the importance of employing copious amounts of water as a rinsing agent in case of exposure, an effective treatment due to the fertilizer's high water solubility.[51] Despite the prudence of such advice, it was most likely not always practical in the often harried conditions of the midwestern farmstead.

EXPERT ADVICE

Safety specialists also identified a number of precautions for using other chemicals, which, if neglected, could result in tragic consequences. Agricultural safety writers provided a wealth of advice, including the need to develop a healthy respect for chemicals. Safety advocates advised farmers to take a cautious approach to chemicals, including properly disposing of empty canisters and rinsing clothing and skin after applying any farm chemicals. They recommended that farmers wear protective devices, follow the label's directions carefully, and remain aware of the larger implications of improper chemical use. The wider repercussions could include inadvertently harming their crops, hurting their neighbors' fields, or even polluting the water resources.

Specialists often believed that the agricultural community had not adequately adopted safe work practices for using chemicals. In 1962, Keith Long, chief of toxicology at the University of Iowa's Institute of Agricultural Medicine, articulated this view. Long stated that farmers were often exposed to hazardous chemicals during both the mixing and the application phases but that their attire was often insufficient, with many of the farmers he observed failing to don any protective apparel or devices while using chemicals.[52] Specialists sometimes recorded shocking details while advocating for safer work practices. For instance, a USDA writer stressed that farmers should "never use [their] mouth to siphon a pesticide from a container."[53] The writers of another chemical safety brochure warned farmers to "never stir any chemical solution with your hands."[54] If based on observation, such practices would have undoubtedly bolstered their argument that current safety practices were insufficient and that expertise was required to solve the chemical safety problem.

Some farmers agreed that safety measures were sometimes inadequate during the chemical application process. While discussing the use of Atrazine on his cornfield, Leroy Hogeland stated that sometimes he "reached in and stirred it with [his] hands," and "sometimes there would be soft chunks in the mix and I'd get in there and mash those chunks up."[55] James Wilson added, "Insecticides were something that we were very, very reckless with. We just took our hands, stirred it around, and then wiped our face."[56] Lawrence Schmitz recalled that if a "sprayer nozzle plugged up, you'd go out there barehanded, take it off, and clean it."[57] Don Rimathe summarized these experiences: "We probably could have done a better job in respecting the dangers associated with chemicals."[58] Although such recollections might appear shocking to contemporary readers, one should always consider the pressures farmers faced to work quickly in often less than ideal conditions.

Safety writers also stressed the crucial nature of rinsing skin and clothing after applying pesticides, particularly prior to smoking or eating. In 1962, Keith Long noted that cleaning procedures varied: "[O]ccasional rinsing in Kerosene may be necessary for clothing with heavy deposits of chlorinated hydrocarbons such as DDT and Dieldrin," but, he continued, "Ordinary washing soda is a better decontaminate for phosphorylated hydrocarbons such as Malathion, Thimet or Parathion."[59] Long's nuanced advice emphasized that different chemicals demanded distinct washing practices, demonstrating yet again the ever-present, ever-expanding complexity of the rapidly industrializing Corn Belt. Some famers followed such advice. For instance, Rex Peterson indicated that washing his skin and clothing was his primary preventative measure in dealing with chemical hazards: "I'd change my clothes as quick as I could when I got done spraying and wash."[60]

Experts also emphasized the importance of wearing proper breathing protection, as well as taking care of such devices. In 1962, Farinholt recommended wearing masks both when applying pesticides and while disposing of used chemical containers. She also stressed that such devices required regular filter replacement, complete and regular cleaning, and proper storage in a "clean dry place."[61] Mining firms manufactured some of the early breathing masks used in agriculture. As early as 1955, a Mine Safety Appliances Company advertisement writer boasted of the "protective benefits" of wearing their brand of breathing devices. According to the writer, an MSA respirator could protect farmers from a variety of well-known agricultural chemicals including: "EPN, Dieldren, Aldrin, TEPP, HETP, OMPA, and Systox."[62]

Although safety writers and manufacturers encouraged farmers to use breathing protection, it appears that farmers initially had little access to such devices. In 1964, USDA entomologists R. A. Fulton, Floyd F. Smith, and Ruth L. Bugsby published their findings in a study entitled *Respiratory Devices for Protection against Certain Pesticides*, which examined the availability and quality of respiratory devices. They found that the development of respiratory equipment for agriculturalists had not kept pace with the rate of chemical innovation, concluding, "No suitable, fully effective respirators were found to be available in 1949 for protecting operators in the field from inhaling dusts or mists of Parathion and related phosphorous insecticides."[63] Farmers were also largely dependent on experts throughout the period regarding the availability of such equipment.[64] It is reasonable to assume that farmers, burdened by time constraints and financial obligations, might have neglected such consultation.

Some safety writers suggested that breathing and other safety equipment was often impractical in the field. Dr. Clyde Berry emphasized the discomfort farmers experienced when using safety equipment, advocating that researchers should don this equipment themselves. He stated, "Wear these protective devices yourselves if you doubt me. Wear them for a full day in the blazing summer sun, and you will return to your air-conditioned office and take another look at the Frankenstein you have created." Berry had also observed that Iowa farmers sometimes dropped their masks to their chins, exposing the mask's interior to air-borne spray residue. Berry noted as well that cleaning small components, such as the spray nozzles, often required the removal of unwieldy gloves, sometimes resulting in direct chemical exposure. He indicated that occasionally the toxic liquid "runs down over the hands and arms of the operator, sometimes reaching the chest and abdomen." Farinholt, too, stated that on occasion agriculturalists used safety devices that were inadequate for meeting the safety requirements of new chemicals. She believed the problem could be corrected by requiring chemical manufacturers to create adequate protective devices for substances before they were allowed to bring them onto the market.[65] Although this might have improved safety, it would have been in opposition to the relatively unregulated environment of chemical use on the nation's farms. Both Berry and

Farinholt's comments suggest that even farmers desiring to follow safe practices while using agricultural chemicals would have often found the effort to be challenging and fraught with difficulty.

PRODUCT LABELS

Agricultural chemical officials and safety specialists persistently emphasized the importance of reading the user guidelines found on labels. In 1955, V. K. Rowe, a Dow Chemical researcher, called attention to the importance of following the label's instructions, expressing his view that if users understood and followed them, "it is doubtful that a single case of ill effects would result from the handling and use of that product. In the interest of safety—read the label!" In 1963, Iowa State Extension writers also advised farmers to carefully analyze labels so as to understand the chemical's properties as well its proper use.[66] Safety writers continually underscored that farmers needed to thoroughly understand a product's label, since the sheer diversity of chemical choices could be dizzying. General categories of pesticides included chlorinated hydrocarbons, organic phosphates, and rodenticides; however, these broad categories encompassed hundreds of individual chemicals. In 1960, J. N. Roney indicated that within categories, chemical groups differed greatly in their toxicity to people and animals, although organic insecticides posed less of a concern than the phosphates. Roney emphasized that the label represented the primary tool to safely navigate the many challenges of such product diversity.[67]

Despite such statements of confidence in labels, substantial evidence suggests that directions were sometimes inadequate. A number of incidents that occurred in the late 1950s and early 1960s illustrate the poor quality of some of the labels. Manuel Velez-Velez and Jamie Ramos-Sanches, two Puerto Rican farmworkers, were applying parathion dust on a Massachusetts vegetable farm. The product label lacked both information regarding the need to dress in protective clothing as well as a skull and crossbones warning emblem; tragically, both men died of parathion poisoning after repeated exposure applying the chemical without protective equipment. On other occasions, individuals received delayed treatment for poisoning when labels lacked antidote instructions. Chemical company officials sometimes composed labels using highly specialized language or excessively minute print, which were generally difficult to read and understand.[68] Even when Corn Belt farm families carefully read and tried to follow a label's recommendations, they remained vulnerable to accidents.

Safety writers maintained it was more common that chemical applicators did not always read the label and that in any event such guidelines could not provide complete information regarding all application situations. In 1964, the writers of a pamphlet cooperatively produced by the NSC, USDA, and the National Agricultural Chemicals Association instructed farmers to "use common sense to bridge the gap between what is written on the label and the actual chemical application."[69] Safety writers also encouraged farmers to confer with a qualified chemical dealer or other authority if the label proved to be insufficient.[70] In 1957, Henry Doyle, an official for the Occupational Health Program, U.S. Public Health Service, stated that unlike manufacturing workplaces, with their controlled settings, those laboring in the farm sector often lacked knowledge of proper safety precautions or sometimes disregarded sound accident prevention advice if they had used other approaches without injury.[71]

Some farmers recognized the difficulty of communicating proper safety precautions within the limited scope of the label format. Jim Klein stated, "How do you explain the chemical composition and warnings to someone without a chemistry background in the space of a label?"[72] Perry Middlesworth believed that labels were difficult to comprehend, especially in the years prior to widespread safety training. He believed that farmers of the mid-twentieth century probably were less precise in applying the correct amount before the availability of detailed chemical application information in later decades.[73] Bernard Stodghill was more concerned with the proper application ratio than with safety information: "The only reason we looked at them was to see how much to mix," he stated, indicating that the specific content and nature of the chemical were of much less concern.[74] Such comments reveal the label's limits in preventing negative health consequences for farmers who lacked agricultural chemical expertise.

Farmers' reliance on product labels underscores the many challenges of chemical safety in the mid-twentieth century. Farmers carried the burden of keeping themselves, their families, and even their neighbors safe from chemical accidents. Although they benefited from an abundance of new chemicals, such substances added tremendous complexities to farming tasks. Unlike manufacturing workers who labored in more supervised settings, farmers were left to educate themselves about proper chemical use. The copious advice given by safety experts regarding the proper mixing, use, and storage of chemicals revealed that it was far from easy. Chemical manufacturers did not always write the labels correctly, suggesting that even reading a product's user guidelines did not guarantee adequate knowledge of the hazards. In addition, both the availability and the practicality of protective devices were not always sufficient, making some safety measures unfeasible. The federal government and the states possessed rather weak chemical regulatory tools prior to the early 1970s, further lessening the possibility of achieving a sound degree of safety. This regulatory vacuum undoubtedly provided farm safety experts with another arena in which to promote their profession, as agricultural work became ever more complex and potentially hazardous.

5

IGNITING THE MOVEMENT

THE ONSET OF THE SECOND WORLD WAR witnessed a rapid intensification of safety efforts aimed at addressing the farm accident problem. The maturation of the necessary institutional tools to coordinate a national effort, labor shortages, and a cultural shift toward professionalization all contributed to the movement's growth. Experts also increasingly relied on statistical methods in investigating problems; this approach served to raise public awareness of previously ignored issues. Although such organizations as the National Safety Council (NSC), state extension services, and rural youth groups existed previously, by the 1940s they had gained considerable influence. The nation's leaders were motivated to support the movement for greater safety by their belief that protecting farm laborers would contribute to victory over the Axis powers. Wartime activities promoting accident prevention were crucial to early organizing efforts, but these efforts continued to mature and evolve beyond the war period. The 1970s saw farm safety activities become more contested and conflicted as legislative, legal, and other new approaches emerged. Yet, despite ongoing measures intended to improve farm safety, agricultural hazards stubbornly lingered into the twenty-first century.

Some researchers have insisted that farmers' switch from using horsepower to using petroleum-powered machines increased their hazards, but little statistical evidence supports such claims. In 1983, Frederick R. Schneider, a retired Farm Security Administration official, indicated that although "specific data" was not available in the early 1940s, he believed that during the mid-twentieth century instances of agricultural injuries and fatalities were on the rise, especially those involving machinery. He indicated that the perception that farm accidents were increasing raised the concern amongst many leaders of the agricultural community that something needed to be done.[1] Mark Aldrich, an expert on the history of industrial safety, concluded that factory hazards lessened after the 1920s, but that risks to farmers most likely intensified as farming became increasingly industrialized. For instance, in 1935, Robert T. Legge, a professor of hygiene at the University of California–Berkeley, stated, "The toll of accidental deaths in agriculture is annually increasing due to the introduction of motor driven farm machinery and electric power." He cited the high accidental death rates in the Corn Belt state of Nebraska in 1929 in support for his contention, but he did not offer any additional data to bolster his argument.[2] The statistical support for such claims is not definitive; however, it is reasonable to assume that the general perception of greater accident frequency helped fuel the development of the farm safety movement.[3]

Historian Theodore Porter provides important insight regarding the role of increasing reliance on statistical methods in the twentieth century in shaping people's understanding of many of society's ills. He argues that the move toward adopting statistics as the "basis" for formulating public policy decisions is crucial to understanding modern life. This movement toward statistical interpretation of reality helps explain experts' increased awareness of the farm accident problem. Porter notes that the "foundations of mathematical statistics were laid between 1890 and 1930, and the principal families of techniques for analyzing numerical data were established during the same period."[4] The farm safety movement emerged after the use of statistics had matured, providing activists with a tool to more fully comprehend the problem. Greater availability of data as well as the perception of headlong industrialization heightened a sense of the severity of the safety issue: Statistical research showcased the unintended consequences of the technological changes that had intensified the hazards of farming.

The safety movement relied upon a network of nonprofit organizations and government agencies, which collectively mounted a voluntary effort to reduce farm accidents. Institutions such as the NSC and the United States Department of Agriculture (USDA) contributed farm accident research activities, created and distributed farm safety information, and assisted in the organization of regional safety programs. State extension employees and farm safety specialists supported and advised the leaders of state and local safety campaigns. Local 4-H chapters, FFA clubs, and a large number of other groups engaged youth in safety activities. As the farm safety effort accelerated, new groups and professions emerged, including state farm safety committees and farm safety experts, both of which were instrumental in supporting ever-diversifying state-level activities.[5] By the 1970s, the farm safety movement had changed in other ways as well. Farm safety efforts were initially characterized by a cooperative atmosphere in which educational awareness campaigns dominated. However, Congress's passage of the Occupational Safety and Health Act, debates about legal accident liability, and disputes within the farm safety advocacy community had replaced the cooperative tone of an earlier era.

A brief review of the history of the major organizations involved in the farm safety movement provides a useful context for understanding the intensification of farm safety activities in the 1940s. This examination is particularly useful since the development of such an effort required a group of vigorous rural-oriented organizations. Both the NSC and the USDA were poised to assume important roles in the farm safety effort. By the 1940s, the leaders of the NSC had already become leading advocates for conducting accident research, establishing local safety committees, and promoting efforts to raise the public's awareness regarding working and living safely. The USDA, land-grant colleges, and state extensions had become the primary means by which the nation's policy-makers attempted to shape agricultural and rural life by promoting modernity and efficiency. In addition, such youth organizations as 4-H and FFA maintained close ties to these organizations, which provided access to the latest agricultural knowledge for the nation's rural youth.

THE NATIONAL SAFETY COUNCIL

Although little had been done in the early decades of the twentieth century to reduce farm accidents, Americans had already taken significant measures to reduce accidents in many other economic sectors. The establishment of the NSC represents one of many actions taken to improve safety in a newly industrialized nation. Manufacturing, like agriculture, had witnessed tremendous technological changes, and, as a consequence, new safety issues had emerged. In 1913, the members of the Association of Iron and Steel Electrical Engineers established the NSC after recognizing the need for an entity devoted exclusively to industrial safety. They divided the organization into sections devoted to improving conditions across a wide swath of industries and occupations. By 1925, the number of the organization's divisions had grown steadily, totaling at that point twenty-six separate sections, covering areas such as refrigeration, cement, and taxicabs, with NSC members devoting themselves to improving safety in their own particular domains.[6]

The NSC emphasized a cooperative and inclusive approach to improving safety, suiting the NSC well to fulfill a crucial role in improving farm safety when very little legislation addressed farm families' working environment. The paucity of coercive, legislative measures to reduce farm accidents helped determine the movement's structure and form. Individuals, businesses, and organizations involved in the movement routinely expressed a cooperative attitude. Intra-organizational coordination was crucial to mounting a successful farm safety program. Agricultural safety advocates often referred to their activities as an effort to "sell safety." This slogan also emphasizes the reality that farm families could not be forced to observe safer work practices: They had to be convinced. In 1913, the NSC's founding members stressed these principles in their commitment "to initiate, promote, cooperate with, and obtain the assistance of any and all activities or agencies calculated to conserve human life." The NSC's cooperative educational approach to accident reduction thus emerged a full thirty years prior to the farm safety movement. The NSC philosophy of cooperative effort to reduce accidents undoubtedly increased the likelihood that they would eventually become instrumental in educational efforts to reduce farm accidents.[7]

The NSC staff fulfilled a variety of key roles in addressing the nation's safety issues, and as a result the organization grew quickly to meet this need. Its functions included conducting accident surveys, establishing local safety councils, and publishing and disseminating educational materials. Royal Meeker, who was appointed the organization's first chair of the Council Committee on accident statistics in 1914, became the U.S. commissioner of labor statistics just two years later. By 1914, the council also had established fourteen local chapters in cities including Chicago, Detroit, and New York, each of which carried out safety campaigns on a local level. By the end of the First World War, the number of local councils had risen to forty-seven.

The NSC also developed the publication services necessary to disseminate safety information nationally. This ideally suited the organization to assist farm safety advocates, who had a great need for agricultural safety materials. NSC members repeatedly emphasized that educating both workers and the general public represented a cornerstone for improved safety. NSC staff members dispersed information through yearly congress transactions, pamphlets, and periodicals such as the *National Safety News*. By 1919, the NSC was also assisting in the development and distribution of industrial and public safety films.[8] The NSC's educational focus most likely explains its intensive role in the early decades of the farm safety movement and its diminished role as the agricultural movement became more litigious and contentious in the 1970s.

Historian Mark Aldrich has indicated that the NSC's educational role represented one of its more important responsibilities, noting that throughout much of its history, "The National Safety Council was most valuable for the service it provided as an information clearinghouse." He pointed out that during the early twentieth century it was difficult for individuals interested in obtaining accident prevention information to collect reliable statistics. However, NSC leadership acted quickly to fill this void by creating a Safe Practices Committee whose members "issued pamphlets on ladders, scaffolds, and other devices, while a bureau of information provided a blizzard of bulletins, posters, movies, speakers, and up-to-date information on changing state regulations." The NSC would assume this role in the farm safety movement by providing the extension staff, farm safety specialists, and youth club leaders with large amounts of literature, research, and programming ideas.[9] The NSC also encouraged farm safety advocates to meet and collaborate with each other. This practice was well suited to the farm safety movement's largely cooperative approach up through the 1950s and 1960s.

During the 1920s and 1930s, the NSC's prominence and influence continued to increase. The organization's annual meeting attracted prominent industrial leaders from a variety of economic sectors, including railroads, mining, and steel. The involvement of such high-ranking businesspeople indicates the organization's national prominence as a leading safety organization. National leaders in the field of safety also regularly attended the organization's annual meeting, including, for example, David Beyer, an early authority on textile safety, who attended the 1917 Congress. Government officials from a variety of agencies, including the Bureau of Mines, state factory inspection departments, and a host of others, attended as well. By

1920, the organization boasted over four thousand members, representing approximately six million workers, and, by the end of the 1930s, it had added a significant number of members from Canada and Great Britain. By the time the United States entered the Second World War, the NSC had become one of the nation's chief advocates for both public and workplace safety.[10]

The NSC began to become more fully involved in the farm safety movement during the Second World War. By 1944, the NSC leadership had transformed the annual farm safety conference meeting into a separate division. Members of the division dedicated themselves exclusively to reducing accidents among farm families and agricultural laborers.[11] The division assumed an important coordinating role in assisting development of the farm safety movement by helping to unify the efforts of other organizations. Committee members also disseminated farm safety materials, assisted in the establishment of state farm safety councils, and engaged in numerous other activities to increase safety awareness. However, central among the NSC's many contributions was providing the necessary synchronization of farm safety efforts. This role was crucial, since as late as 1940 individual states conducted much of their farm safety awareness activities without interstate cooperation.[12]

A highly competent staff, including individuals who had previously held respected positions in agricultural organizations, bolstered NSC's activities to reduce farm accidents. Such institutions included the USDA, the farm press, land-grant colleges, and the state extension services. Maynard Coe's appointment as director of the farm division in the winter of 1944 confirmed the NSC's wise hiring practices. Coe possessed impressive agricultural credentials, including having served as Kansas's 4-H club state leader. C. L. Hamilton, the division's assistant director and editor of the *Farm Safety Review*, possessed a master's degree in agricultural engineering from Iowa State College; prior to his NSC appointment he had held a faculty position at the University of Saskatchewan and had served as field inspector for the Soil Conservation Service throughout the 1930s. Individuals with agricultural engineering backgrounds would continue to dominate the NSC's farm division leadership and staff positions,[13] ensuring that NSC farm safety activities were firmly in the hands of agricultural experts and reformers.

The NSC's efforts to facilitate the development of a network of organizations in the farm safety movement also depended on generous private-sector financial support. The NSC's adeptness in fund-raising is demonstrated in the *Farm Safety Review* benefactor lists. The numbers and diversity of companies providing monetary support expanded throughout the era. In 1946, a *Farm Safety Review* writer mentioned ten companies and organizations, primarily agricultural in nature, providing funding for the NSC's farm division, including the John Deere Company, the International Harvester Company, and the Illinois Agricultural Association. During the 1950s, the number of sponsors continued to grow: A 1960 *Farm Safety Review* article disclosed the names of 196 supporters. This more varied group of businesses included insurance companies, steel companies, and chemical firms. In 1963, Howard Pyle, president of the NSC, acknowledged the importance of these private contributions, stating, "We salute the individuals and organizations whose direct financial support provides convincing proof of their interest in the advancement of farm safety efforts."[14] Such funding was necessary for educational materials development, research activities, and administrative leadership. The NSC even depended upon private contributions to fund the *Farm Safety Review*'s publication and distribution. The *Review* represented the only periodical aimed exclusively at serving as a mouthpiece for the farm safety movement.[15] This financial support was crucial in the NSC's development of its key awareness campaigns, such as Farm Safety Week.

This annual safety awareness campaign illustrates one of the many ways the NSC assisted in creating cooperative and effective farm safety programs. In 1944, the NSC's farm division members appointed a five-member board to develop the first Farm Safety Week. These administrators received advice from USDA officials who also provided assistance in the program's implementation. In addition, both organizations cooperated in the development, distribution, and preparation of safety educational materials. President Roosevelt underscored the importance of these efforts both for experienced farmers and for the thousands of new workers engaged in agricultural work. Each day during Farm Safety Week was devoted to a specific farm hazard. Farm safety specialists, extension agents, and rural youth club leaders directed efforts such as tractor-tipping demonstrations, farm hazard inspections, and a variety of other activities aimed at increasing safety awareness.[16]

The success of the first Farm Safety Week illustrated the communicative powers inherent in the farm safety movement's organizational network. The NSC appointed chairs in forty-five of the then forty-eight states to direct that state's Farm Safety Week activities. These state leaders included representatives from the farm press, extension service directors, and university professors. The NSC and the USDA sent educational materials and program guides to these leaders, who then distributed them to their agricultural organization members. E. W. Lehman, chairman of the NSC Farm Safety Week Committee and University of Illinois Agricultural Engineer, highlighted the event's intra-organizational collaboration, stressing that hundreds of organizations had participated and that the NSC had played a vital role in publicizing Farm Safety Week across the nation.[17] The participation of individuals such as Professor Lehman symbolized the level of involvement of experts who wished to alter farm families' approach to their work. Throughout the late 1940s, the leaders of such organizations as the NSC, USDA, the Farm Bureau, the National Farmers' Union, and the National Grange repeatedly emphasized the need to assist each other, and they joined in observing Farm Safety Week as an expression of the collective will to reduce farm accidents.[18]

National Farm Safety Week continued to be an important tool for raising rural safety awareness, gaining increased participation and support throughout the era, including additional corporate and media backing. In 1951, Coca-Cola incorporated the farm safety message in its national advertising campaigns, and both NBC and ABC addressed farm safety concerns in their rural-oriented programming, which included shows such as NBC's *National Farm and Home Hour* and ABC's *American Farmer*. The NSC also distributed increasing quantities of safety material to rural educators, supporting associated local activities. In 1955, state extension leaders received ten thousand Farm Safety Week information packets, and local safety leaders led five thousand local safety meetings.[19] In 1962, a *Farm Safety Review* writer stated, "More than a million pieces of material will go to every county and to the rural press and radio this year. Every state and its farm organizations will participate in Farm Safety Week in an effort to make every farm family safety conscious fifty-two weeks of the year."[20] In 1965, NSC employee Jack Burke indicated that, for safety advocates across the nation, Farm Safety Week signifies a "high point" in their yearly calendar of events.[21]

The NSC's contributions in enlisting and assisting youth groups such as 4-H and FFA in pursuing its safety activities represented another important contribution to the effort to reduce farm injuries and fatalities. NSC hosted an annual youth farm safety sectional meeting, demonstrating the organization's commitment to expanding the involvement of FFA, 4-H, and other youth-oriented organizations. In 1952, NSC leaders inaugurated the annual youth meeting as part of the National Safety Congress convention. In 1955, 130 "advisors and delegates" represented organizations such as 4-H, FFA, and the Farm Bureau Young People.[22] The number of "young people and adult leaders" present at the meeting expanded to two hundred the following year.[23] By 1959, "more than two hundred youth and seventy-five adults attended the Rural Young People and Farm Adult Sessions," held at Chicago's Palmer House Hotel. The conference consistently provided a venue for concerned leaders and youth involved in safety efforts. Participants gained the latest safety training, received recognition for their achievements, and met other youth involved in safety promotion.[24]

The NSC adopted an inclusive approach both in terms of planning the youth sectional meeting and of encouraging the participation of a diverse group of organizations. Representatives from 4-H, FFA, the American Farm Bureau Young People, Grange Young People, as well as many other groups attended the youth sectional meeting. The national FFA, Farm Bureau, and 4-H adult advisors assisted NSC staff members in planning the conference activities with the youth. For instance, NSC employees, several outstanding youth club members, and their adult leaders planned the 1955 meeting. A *Farm Safety Review* writer reflected on preparation for the event. The author stated, "Plans for this year's program include the presentation of exhibits and demonstrations, reports on outstanding local youth safety programs and specific safety projects, outstanding youth highway safety programs, workshop sessions on power tool safety in the home farm shop, and 'buzz' (informal discussion) sessions."[25] Organizers of the event also adopted a yearly theme; in 1956, they adopted the motif of "Let's plant, cultivate and harvest safety"; the following year, organizers proclaimed "Youth Backs the Attack."[26]

Rural youth attested to the benefits of the annual youth gathering to the promotion of safety in their communities. Participants indicated that they learned valuable safety information, received recognition for their achievements, and socialized with others concerned about accidents. Safety speeches were delivered by national 4-H and FFA leaders, tractor safety was demonstrated, and a host of other educational events were offered. Attendees also enjoyed formal dinners, dances, and other entertainments. Speakers included members of prominent safety organizations, corporate officials, and safety experts. In 1956, session titles included "Fire Drills for Farm Families," "Fire Demonstrations," "Visual Aids or Gadgets That Can Teach Safety," "First Aid," "Water Safety," "Corn Picker and Tractor Safety," "Gun Safety," and "Falls." The annual Farm Safety Youth Section usually also included award ceremonies and tours of local attractions.[27] Thus, these NSC gatherings helped create an interdependent web of relationships connecting a variety of government agencies, farm organizations, and other agricultural interests, forming a cooperative network focused on decreasing the dangers present on the American farms.

THE USDA

The NSC was not alone in its effort to meet the challenge of the farm accident problem, as the success of Farm Safety Week demonstrated. Federal agencies such as the USDA and state institutions such as land-grant colleges and extension agencies fulfilled key roles in the farm safety movement. All of these organizations had been in existence decades prior to the emergence of farm safety efforts; however, by the eve of World War II, their ability to assist in mounting a farm safety effort had grown. The USDA had developed from modest beginnings into a potent tool for federal agricultural policy making. Land-grant colleges, state extension programs, and rural youth organizations had also evolved into powerful tools for agricultural education at the state and local levels.

By the early twentieth century, the USDA had demonstrated that it was the primary instrument by which the federal government shaped agricultural policy. The USDA had grown from a small organization during the Civil War to one wielding considerable power. The legislation that formed the USDA delineated functions that eventually led to engagement in farm safety efforts. These institutional aims included providing agricultural education to the American public, conducting and disseminating the results of crop and animal research, and instigating statistical studies, all of which strengthened the organization's role in the safety movement. Initially, the USDA limited its responsibilities to increasing farmers' productive capacities. However, as the decades progressed, the USDA's roles extended significantly to include efforts to improve the healthfulness of the American food supply, expand the nation's rural transportation infrastructure, and administer a diverse array of farm subsidy programs.[28]

The USDA not only helped in collective efforts to raise awareness about efforts to reduce farm accidents, it also engaged in farm accident research projects and promoted national meetings encouraging collective action.[29] Thus in many ways the USDA fulfilled a role similar to that of the NSC in the farm safety movement by fostering intra-organizational coordination, information dissemination, and research. For instance, in 1944, the USDA established a Safety Council whose members studied the issue and composed educational materials for public distribution.[30] Agency officials also helped plan the Occupational Safety Conference held in May 1956 in Washington, D.C. The three-day event included a section on the issue of safety in agriculture. Sam Lyle, who served as the USDA's assistant director of agricultural programs, assisted the extension service in coordinating the meeting, which brought together farm safety leaders from throughout the nation and resulted in specific recommendations to be given to the president.[31]

Although the USDA did not initially play a large role in the farm safety movement, the agency intensified its safety-related work in the early 1970s. For instance, the organization established a more active Farm Safety Committee. The USDA's Task Force on Safety in Agriculture symbolized a renewed commitment to reducing farm accidents. The department's enlarged role was prompted by the increasingly litigious and regulatory environment of the 1970s. In 1971, Lloyd H. Davis, a high-ranking USDA official, summarized the reasons behind the department's enlarged role, indicating that although USDA employees had studied the safety issue and provided educational material for decades, passage of new legislation such as the Occupational Safety and Health Act in 1970 justified an enlarged effort. Davis believed the expertise and knowledge of USDA officials would be crucial in the increasingly regulatory environment of the late twentieth century since their expertise would be necessary to develop new standards and regulations.[32] The 1970s also saw the USDA begin providing funds to assist state farm safety programs, including the hiring of farm safety specialists.[33]

THE LAND-GRANT COLLEGE AND EXTENSION SYSTEMS

By the time of the Second World War, land-grant colleges had developed considerably from their nineteenth-century origins. Late in the nineteenth century, leaders of these institutions had initially struggled to establish a set curriculum. Over the ensuing decades they had greatly clarified both their missions and their course offerings, developing into highly regarded and influential institutions of higher education. By 1940, sixty-seven land-grant universities were training hundreds of individuals in a variety of fields with a profound impact on agriculture. These colleges also housed the nation's agricultural education training programs and the expanded agricultural engineering programs that produced Vo-Ag teachers and agricultural specialists of every variety.[34] These teachers and specialists, in turn, directed rural youth organizations and filled the ranks of the state extension services.[35]

State extension had been in existence for decades prior to the Second World War. With support from the Smith Lever Act of 1914, state extensions received federal support. Agricultural scholar Willard Cochrane grasped extension's significance, stressing that its creation had "institutionalized the outreach programs of the state agricultural college." He also emphasized the interconnectedness of land-grant colleges and extension services, describing extension personnel as basically "off-campus instructors of the colleges of agriculture."[36] Historian David Danbom recounts that USDA officials, land-grant college faculty, and experiment station employees sought more effective ways of delivering the latest agricultural knowledge into the countryside. As a consequence, they worked together to accelerate the development of the extension service in the belief that the extension system would intensify the pace of agricultural reform by increasing farm families' efficiency and productivity.[37] Despite such lofty goals, the extension agents faced various challenges throughout the early decades of the twentieth century.

Historian Dorothy Schwieder noted that early agents both lacked an institutional support structure and, in some cases, faced a skeptical audience.[38] Some rural residents resented the program, believing its existence implied that farmers were incompetent and needed outside help.[39] Nevertheless, state extension programs continued to expand and gain strength throughout the early twentieth century. During the First World War, extension's importance increased as it was enlisted as the primary institutional tool for mobilizing the nation's agricultural sector. Extension officials assisted in efforts aimed at meeting labor shortages, overcoming agricultural supply shortages, and increasing food production. Historian Wayne Rasmussen indicated that World War I represented the first time extension had been used to meet the needs brought about by a nationwide emergency.[40]

The nation's extension programs continued to expand markedly between the 1920s and the 1950s. The demands of the Great Depression, environmental stress, and wartime emergency produced a need for larger staff, expanded publication services, and new communication tools, including radio.[41] Thus, by the onset of the Second World War, extension programs had been transformed from their beginnings into a vigorous rural education network. State extension leaders recognized the potential power they could offer and worked to increase communication capabilities. Ralph K. Bliss, Iowa's director of extension, stated that with more work "it is possible to supply information to nearly every resident in Iowa within a week from the time it is started from the administration at Washington." Bliss's comments indicate how effective the communicative and bureaucratic infrastructure of extension had become, allowing a myriad of agricultural education efforts and wartime mobilization activities to be initiated and carried through. His claim was clearly validated by the extension service's effectiveness at meeting the wartime agricultural labor shortage.[42] At the same time, this communication network proved indispensable in raising awareness about the farm accident problem during the Second World War.

The Cooperative Extension System served as the primary means by which expertise and agricultural knowledge were dispersed into the countryside throughout the twentieth century. Cooperative Extension employees and their reformist allies labored to disseminate a diverse array of "improvement" and "modernization" messages to farm families.[43] Their efforts to update agriculture included promoting such ideas as "efficiency," "order," and a reverence for "expert knowledge."[44] Such values manifested themselves into a diverse array of measures to help meet rising food demands, including promotion of school consolidation, encouraging rural women to adopt "modern household equipment," and promoting the latest agricultural methods.[45] These principles promoted improved agricultural productive capacity as well as farm accident reduction.

Extension workers aimed their efforts at activities both inside and beyond the farmhouse. Female home demonstration agents largely communicated information relating to "such domestic skills as cooking, canning, clothing construction, and child rearing while farm agents with information on crops and animals worked exclusively with men."[46] Despite the extension service's gendered view of farmwork, in reality distribution of responsibilities on farms was greatly more complex. Scholar Katherine Jellison emphasized that extension policy in the early years overlooked "that much of farm women's work took place outside the farmhouse kitchen. In short, they did not perceive the interdependence and fuzzy distinction between the farmhouse and the farm itself." For instance, farm women labored in gardens, chicken coops, barns milking cows, and a variety of other agricultural settings and tasks.[47] Their limited concept of women's roles supports the view that extension employees did not always fully understand the circumstances of the people they served.

State extension agencies played an important part in the farm safety movement by participating as an important constituent in the network of cooperating organizations. Extension leaders, as did the leaders of other agencies involved in the effort, believed cooperation to be crucial for progress. In 1946, M. L. Wilson, the national director of Cooperative Extension, emphasized the need for state extension workers to assist other organizations in a common effort to improve rural life. Wilson stated, "We need to work closer with, not shy away from, farm organizations and other cooperating groups in those activities directed to attainment of higher living standards and community welfare."[48] This cooperative approach was consistently demonstrated in the way farm extension agents assisted youth clubs in safety activities, worked with other safety specialists, and provided leadership for a variety of safety programs.

Home demonstration agents also proved to be important assets, exemplifying community engagement. These dedicated women instigated a variety of safety programs and campaigns on diverse issues, including the rural auto accident problem, farm home safety, and children's well-being on the farm. They initiated and participated in farm safety efforts in Indiana, Iowa, and many other Corn Belt states as early as the 1940s. Home demonstration leaders also encouraged extension agents to integrate safety into their other activities. In 1959, the National Home Demonstration Council, an association for such professionals, outlined an effort directing its members to reduce accidents in their communities. The plan called for home demonstration agents to educate themselves regarding local issues, to instigate safety programs, and to work with other organizations already involved in the movement.[49]

4-H AND FFA

The rural educational infrastructure was further enhanced by the expansion of 4-H. Although the youth organization had modest beginnings as locally organized youth agricultural clubs in states such as Ohio, Illinois, and Iowa, the support of both the USDA and land-grant colleges greatly assisted its growth. The leaders of these institutions quickly recognized the capacity of these youth groups to disseminate the latest agricultural information to farm families. The Smith Lever Act helped formalize the close affiliation between the nation's extension officials and its 4-H clubs and made available a continual stream of funds from the national government to support 4-H club development.[50]

The continued expansion of 4-H during the interwar years was due not only to the organization's close affiliation with extension, however, but also to growing private-sector support. In 1921, the creation of the National Committee on Boys and Club Work greatly assisted the organization in obtaining such financial backing. Committee members included representatives of large food processing companies, various agricultural concerns, and even department stores. Historians Thomas Wessel and Marilyn Wessel, historians of the 4-H movement, stressed the importance of the committee's work in attracting the private benefactors indispensable for expanding 4-H programing. The committee not only worked to provide funds for 4-H work, it also established an annual 4-H club congress to serve as a national meeting place for the 4-H leadership and youths. Youth Club Farm Safety activities also benefited greatly from such support.[51]

Growth of 4-H accelerated during the Second World War and continued well beyond the conflict's conclusion. The organization's increasing profile was reflected in the growing numbers of rural youth enrolled in 4-H club chapters, the diversification of educational programming, and the group's interconnectedness with other national organizations. Membership in 4-H clubs expanded from approximately 800,000 in 1929 to 1,500,000 by 1940 and approximately 2,000,000 by 1950. The group provided these many thousands of young people with a variety of educational opportunities. Events such as 4-H's annual National Congress and National 4-H Club Week also assisted in increasing the organization's prominence. By the mid-1950s, 4-H chapters had expanded across the United States and to forty-five other countries.[52]

The interest of the nation's agricultural leaders and sustained governmental support also aided the expansion of the Future Farmers organization. State and local lawmakers expanded vocational agriculture programs in the nation's high schools, and federal authorities accelerated their development through legislative means. The Smith-Hughes Vocational Education Act of 1917 hastened such expansion, devoting federal funds to vocational agricultural (Vo-Ag) programs. Such financial support consequently assisted local Vo-Ag clubs connected to programs in vocational agricultural educational. In 1926, national Vo-Ag leaders met in Kansas City, Missouri, for a national congress of Vo-Ag students. In the fall of 1928, leaders of agricultural education, students, and Vo-Ag instructors moved to establish a national FFA organization at the meeting.[53]

The Future Farmers of America achieved greater prominence and influence in the years just prior to intensification of efforts by the farm safety movement. FFA leaders established a national headquarters, gained support from advocacy groups, and initiated national publications. In 1939, FFA leaders initiated construction of a national headquarters on the former grounds of George Washington's estate. In 1944, the National FFA was established, representing the organization's success in gaining corporate and individual financial support. These donations were crucial in providing funds for an ever-growing array of FFA programs, including those concerned with farm safety.[54] In 1979, Ralph E. Bender, a professor of agricultural education at the Ohio State University, assessed the importance of the foundation, finding that it continually served as "an opportunity for business and industrial concerns and individuals to make contributions to support the activities of FFA at the national, state or local level."[55] FFA membership also continually expanded throughout the mid-twentieth century. In Iowa alone, for example, FFA membership grew from 3,664 in 1930 to 7,900 in 1940.[56]

WARTIME MOBILIZATION AND THE FARM SAFETY MOVEMENT

The farm safety movement emerged not only in response to impetus from national organizations, but also, just as importantly, from the political context of the Second World War. This global conflict greatly accelerated a nascent recognition of farm dangers, turning the farm safety movement into a coordinated effort to reduce farm accidents. The nation's political leaders understood that accidents wasted precious human resources needed for the war effort. Responding to the Japanese attack on Pearl Harbor, the U.S. government initiated mobilization efforts in many areas of American life. President Franklin Roosevelt and his advisors saw that the Axis powers' considerable military might, dramatic territorial acquisitions, and totalitarian political systems threatened our nation's very existence. The federal government expressed its zeal for victory by creating new organizations aimed at structuring the U.S. society and economy to meet the Axis threat. Congress passed the War Powers Act, providing the president with tremendous power and giving rise to such agencies as the War Production Board (WPB), the War Manpower Commission (WMC), and the Office of Scientific Research and Development (ORSD). These institutions provided the nation with the bureaucratic apparatus necessary to defeat the Axis powers.[57]

The nation's leaders also recognized the importance of marshalling food producers to achieve sufficient production to meet the demands of a global conflict. In January 1943, President Roosevelt emphasized the significance of food production while delivering a speech encouraging Americans to continue their wartime sacrifices. He stated that "food is the life line of the forces that fight for freedom. Free people everywhere can be grateful to the farm families that are making victory possible."[58] His secretary of agriculture, Claude Wickard, also emphasized agriculture's vital role, stating, "Food will win the war and write the peace."[59] Both Roosevelt's and Wickard's comments reveal the importance of concerted efforts to increase production to feed American troops, supply allied armies, and sustain the populations of liberated countries. These measures included a vigorous publicity campaign to ensure an adequate food supply. The Office of War Information and the War Food Administration saturated Americans with a continual stream of posters, pamphlets, and film clips stressing the need to meet production goals.[60]

Americans displayed patriotism and cooperation during the war in many ways, including the effort to increase food production capacities. The president echoed these nationalistic sentiments when addressing the nation's 4-H clubs in 1943. President Roosevelt congratulated the approximately one-and-a-half-million youth who had participated in the 4-H Victory Program. This plan included an entire range of activities aimed at enlisting rural youth to aid in the mobilization effort. Roosevelt encouraged 4-H members to continue their efforts in a letter sent to 4-H clubs. He stated, "Let your head, heart, hands, and health, truly be dedicated to your country, which needs them now as never before."[61] In 1943, a United States Army representative congratulated Michigan farmers for greatly increasing bean and sugar beet production. This appreciation culminated in a week-long event that included various exhibits of military equipment and speeches encouraging farmers to continue their productivity drive.[62]

Government officials and agricultural leaders informed farmworkers, rural youth, and even agricultural engineers that their efforts were as important as the efforts of soldiers in the military theaters. Agricultural officials often stressed the importance of farmers' responsibilities in winning the production battle. The very names given to rural mobilization campaigns symbolize the nationalism infused into agricultural work. The titles of programs such as the "Women's Land Army," "Food for Freedom Campaign," and the "4-H Victory Program" all signified that food production represented a battle requiring organization, national loyalty, and cooperation.[63] H. B. Walker, president of the American Society of Agricultural Engineers, stated that "one year ago today a treacherous enemy plunged our nation into a war. We choose now to call it a global war in which we have two distinct fronts. One is a military front and the other is a food front." He also stressed that agricultural engineers had a special responsibility in helping to meet the enormous food demands of the war.[64] It was

Walker's occupational group who would be charged with mounting professional efforts to reduce the waste associated with farm accidents.

The nation's leaders quickly discovered potentially serious obstacles to achieving victory in the production battle. These challenges included the problem of labor scarcity, as farmworkers migrated to cities and men enlisted into military service. A study by the Bureau of Agricultural Economics highlighted the dilemma, reporting that approximately two million men had been diverted from agricultural work to other employment or military service. In 1944, a USDA report revealed that approximately eight hundred thousand women and 1.2 million young people would be required to meet food production goals. The national press, including both *Time* and the *Washington Post,* featured articles stressing the need to solve this predicament.[65] Wesley Longnecker, a lifetime farmer from Polk County, Iowa, remembered, "We ran that corn picker almost continuously" during harvest season due to labor shortages during the Second World War. Longnecker's comments reinforce the point that production goals had concrete consequences, including speeding up the pace of work and increasing the chances of accidents while harvesting.[66]

The public responded to the labor problem by participating in a host of programs, including the Victory Farm Volunteers and the Women's Land Army (WLA). During the war, roughly 2.5 million young people joined the Victory Farm Volunteers and over 1.5 million women participated in the WLA. The WLA, for example, had been created by the nation's agricultural leaders in 1943 to help remedy labor shortages in the food production sector. Nevertheless, solving the labor issue required persistent effort. John R. Fitzsimons, Iowa supervisor of farm labor, said that his state had "scraped the bottom of the hired man field" in order to fill the jobs in such shortage areas as canning facilities, small grain harvest, and corn detasseling. Nevertheless, during the summer of 1945, Iowa's farm labor offices had successfully placed approximately seventy thousand workers.[67]

Agricultural leaders also argued convincingly for intensified efforts to reduce farm accidents during the war. Accidents represented a waste of human resources. Labor shortages meant the enlistment of inexperienced workers in the agricultural workforce, and reducing accidents represented a moral cause. In the spring of 1944, Maynard Coe, director of the NSC's farm division, offered such justifications stating that "an estimated four million extra farmworkers will be needed this year.

Many of these workers will be inexperienced in farming, so they will present an unusual safety problem. Farm labor leaders recognize this fact and are working hard to provide adequate training, good supervision, and safe working conditions. They know that the efforts of many workers are nullified by the property and time lost due to accidents."[68] Coe's professional background in 4-H and higher education gave his views the wide credence needed to help bring together the institutions and organizations that could make such voluntary efforts possible.[69]

The members of the emerging agricultural safety movement also stressed a sense of national duty, patriotism, and cooperation, and they believed that reducing accidents should be a national priority. A variety of groups, including agricultural engineers, members of the National Safety Council, and a host of other individuals, stressed that safety in farmwork was a patriotic duty. E. W. Lehman, a University of Illinois faculty member in agricultural engineering, and H. P. Bateman, a member of the American Society of Agricultural Engineers, emphasized the importance of establishing a safety program as part of the mobilization effort. They indicated that rural and fire safety programs were essential, citing the scarcity of both skilled labor and machines. H. L. Miner, chairman of the National Safety Council's Farm Safety Committee, stressed that everyone should work to eliminate farm hazards with the same tenacity with which American soldiers were fighting overseas.[70] Some concerned individuals even considered workplace negligence to be an unpatriotic act. Harold H. Beaty, a member of the Iowa State College faculty, stated that "the careless farmer who gets injured in an accident this year not only hurts himself and his family but curtails the nation's food for freedom campaign."[71] George M. Gehant, an extension agent from Yellow Medicine County, Minnesota, concurred, stating that lost time due to accident could have been "used to good advantage producing food, fiber, fats and other materials needed by the boys in our fighting forces."[72]

The nation's political leaders and those directly involved with solving the labor problem emphasized the importance of wartime farm safety efforts, understanding that thousands of inexperienced farmworkers needed safety education. President Roosevelt expressed the need for reducing farm hazards while issuing a National Farm Safety Week proclamation. He indicated that safety was especially important given the national problem of thousands of unseasoned laborers entering the agricultural sector for the first time.[73] Claude R. Wickard, secretary of agriculture, also indicated that the increase of urban workers entering the farm labor force represented a potential hazard. Florence L. Hall, chief of the WLA, noted that thousands of unseasoned female agricultural workers were entering the ranks of the farm labor force. The individuals responsible for these thousands of untested workers were quick to recognize the great number of farm-work hazards they faced, ranging from farm machinery to difficulties in securing a safe water supply.[74]

STATE FARM SAFETY COMMITTEES

State farm safety committees were crucial for the implementation of efforts to reduce accidents at the state level. They fulfilled a variety of purposes, including cooperating with other agricultural organizations to raise safety awareness and hosting meetings to provide opportunities for exchanging ideas and formulating plans. They also tailored programs to suit the unique needs of different agricultural regions and were indispensable in furthering the involvement of state government agencies, local 4-H clubs, and Farm Bureaus in the safety cause. In 1983, Herb Plambeck, a longtime member of the Iowa Farm Safety Council, listed forty-nine accomplishments his committee had achieved over its forty-year existence. These included continually cooperating with both the NSC and the USDA, enthusiastically participating in Farm Safety Week, and conducting farm accident studies. Plambeck also listed numerous activities the council had supported in conjunction with the state farm safety specialist, various extension staff, and youth groups. Farm safety councils and committees in other states fulfilled a similar role and made comparable contributions.[75]

The NSC played an important advisory role in the creation of committees in states such as Indiana, Oregon, and Washington. For example, in 1942, members of both the Indianapolis Safety Council and the NSC's farm safety division cooperated in organizing a meeting of many of the state's prominent rural organizations. Individuals representing such groups as the Grange, Purdue University, and the Indiana Farm Bureau met to establish a state farm committee. A writer for *Farm Safety Review* stated that all attendants were "deeply concerned at the appalling loss of life and limb and property as the nation faced the terrific impact of World War II" and that "there was general agreement that a coordinated attack on farm accidents in Indiana was long overdue." The attendees responded by creating the Indiana Farm Safety Committee to provide leadership and coordinate efforts to reduce the numbers of farm accidents in the state.[76] The NSC provided similar assistance with both Oregon's and Washington state's farm safety committees. In 1960, an anonymous *Farm Safety Review* writer stated that one of the major duties of the organization's farm division was to provide "assistance to the 45 permanently organized State Farm Safety Committees" and to continue the work of "organizing new State Farm Safety Committees."[77]

The numbers of farm safety committees expanded quickly during the movement's first two decades, from a mere three in 1944 to thirty-one by 1955. By 1960, safety advocates in forty-five states, including members of the farm press, insurance companies, government, and agricultural-related agencies, had formed such committees.[78] For instance, in 1943, Iowa farm safety advocates formed one of the nation's first farm safety councils. In 1983, while commemorating the accomplishments of Iowa's Farm Safety Council, Herb Plambeck, a longtime agricultural broadcaster for the WHO radio station in Des Moines, stated, "Four men are commonly recognized as the earliest pioneers in organized farm safety work in Iowa. They are the late Kirk Fox, editor for *Successful Farming*; the late J. S. Russell, farm editor for the *Des Moines Register*; B. W. Lodwick, a longtime leader in Iowa; the National Farmer's Home Administration; and L. J. Long Keeny, president of the Grinnell Mutual Reinsurance Companies."[79] Agricultural engineers and others employed at the nation's land-grant colleges were also heavily represented on these committees, thus reconfirming the great influence such experts had in shaping the movement. The farm safety committees in other states were founded by individuals with comparable backgrounds and interests.[80]

By 1960, almost every state in the nation possessed a farm safety committee, which reveals that farm safety had been institutionalized as a field of action and knowledge. In addition, once the members of this network of organizations had recognized the problem and enlisted a commitment from its members, it proved to be a potent educational force. Safety advocates demonstrated their power to mount safety education on the national, state, and local level. Farm safety participants exhibited high levels of cooperation throughout the era by pooling their collective resources to raise safety awareness regarding almost every conceivable farm-related health risk threatening the farm family.

INSTITUTE OF AGRICULTURAL MEDICINE

The Institute of Agricultural Medicine represented another important element in addressing the farm accident problem. The organization cooperated with other farm safety organizations while also pioneering new methods to improve agricultural health. In 1955, a grant from the W. K. Kellogg Foundation was instrumental in funding the institute's establishment. Dr. Richard A. Tjalma, an assistant professor of hygiene and preventative medicine of the University of Iowa's College of Medicine, stressed the organization's unique focus, noting that institute staff members were dedicated to the "demonstration and evaluation of public health problems of a rural nature." By 1958, the institute also exhibited an impressive degree of professional diversity, including "a medical director, an industrial hygienist, a social anthropologist, a toxicologist, a statistician, and a public health veterinarian." Although the agency was housed within the College of Medicine at the University of Iowa, during the 1950s its staff established a precedent of collaboration by cooperating with the Iowa State Extension Service, the Iowa Department of Health, and the Iowa Farm Bureau federation.[81]

The institute's activities were widespread and varied during the late 1950s and early 1960s. Such efforts ranged from mounting public awareness campaigns to conducting farm accident studies. Institute staff encouraged rural families to adopt preventative health measures, including immunizations. They also devoted their energies toward such varied efforts as ascertaining the threat posed by such zoonotic diseases as leptospirosis, psittacosis, and rabies.[82] Additionally, they examined individual farm accidents to ascertain common factors that led to farm deaths and injuries. Such investigations included collaborating with others in efforts to not only identify accident causes but also work towards preventative measures. In March 1963, L. W. Knapp, a key member of the institute, corresponded with Norval Wardle as both men studied a fatal silo accident in Sheffield, Iowa. Such examples illustrate the collaborative spirit with the diverse institutions and organizations making up the farm safety movement.[83]

Institute specialists also worked to enhance their understanding of the hazards presented by agricultural machinery. This effort led to the creation of the institute's accident prevention section. These activities intensified when the Kellogg Foundation provided the organization with a $59,000 grant to bolster efforts to address mechanical hazards. Researchers labored to improve statistics surrounding agricultural equipment accidents and to disseminate their findings to the larger agricultural community.[84] In 1968, an article appeared in the *Iowa City Press Citizen* listing the many areas of research conducted by the institute's accident prevention section. These included "farm tractor–motor vehicle accidents, eye injuries, the effects of tractor noise on hearing, safety shielding with corn pickers, tractor anti-roll frames, design of rear guards, shielding for rotary power lawn mowers, a six-county rural family accident survey, and an extensive study of power take-off injuries." In addition to support from the Kellogg Foundation, such activities received support from the U.S Public Health Service and the Architectural Aluminum Manufacturers Association.[85] Thus, the institute represented a potent marriage of public expertise coupled with a continued flow of private resources.

In 1966, the institute's mission to address agricultural health problems was furthered by the addition of a new facility. The Kellogg Foundation provided a grant of $223,491 for the erection of a new building featuring classrooms, a large lab to test equipment, and staff offices.[86] The facility was located north of Iowa City on the grounds of the Oakdale Sanatorium. This tuberculosis treatment center had originally opened in 1908 but the number of patients had shrunk by the 1960s. This positive development had led University of Iowa administrators to repurpose the 560-acre site into a vibrant medical research and treatment center addressing a wider range of public health concerns, including agricultural health.[87] L. W. Knapp, a safety engineer and the director of accident prevention at the institute, celebrated the opening of the new facility, stressing, "The laboratory will give us an area in which we can work closely with farmers and farm machinery manufacturers who want to improve the safety of farm equipment."[88] Such efforts quickly enlisted the support of many of the nation's farm equipment manufacturers, who provided funding to support new research activities.[89]

Such physical expansions were paralleled by the organization's continued involvement in the latest accident prevention approaches. For instance, as early as 1964, Knapp encouraged members of the farm safety movement to alter their focus. He stated, "We have created and perpetuated the idea in everyone's mind that accidents are an individual's own fault, neglecting the things that can be done mechanically and physically." He implied that "greater consideration" should be given to modifying both farm equipment and the environment to better suit operator needs. He contrasted this approach with earlier efforts to modify human behavior to prevent farm accidents and cited experiences in several other industries, such as washing machines and glass production, in which engineering improvements had dramatically reduced injuries. For instance, glass manufacturers had begun making "a glass that, when it did break, would crumble instead of splinter."[90] Knapp's comments foreshadowed the intense discussion about tractor rollover protection devices, which would become more common in the 1970s.

Institute researchers had also begun to examine the possibilities for adopting a rehabilitative approach to assist those who faced the lingering effects of farm injuries. In 1966, Knapp indicated that institute researchers would enlist the help of doctors at the University of Iowa to help address this neglected issue. He stated, "We will work with the department of orthopedics at the University of Iowa College of Medicine and the local county rehabilitation offices to design controls to fit the handicap of the farmer." He emphasized, "In this way we hope to overcome the increased risk of accidents a handicapped farmer experiences when he operates farm equipment."[91] Such services had been stressed for military veterans since the First World War; however, little attention had been given to this approach previously in the realm of farm safety.[92]

The institute's leadership included prominent members of the midwestern farm safety movement, who also developed national and international reputations. In 1968, Dr. Clyde Berry, who at the time was the institute's associate director, was a member of the Air Pollution Control Agency's Development Advisory Committee. The group provided consultation with the National Air Pollution Control Administration in its efforts to improve national air quality. Nevertheless, as shown by articles he wrote for such well-known publications as the *American Journal of Public Health*, Berry continued to work to educate the public about the unique health conditions of those who labored in agriculture.

Other members of the institute had similar accomplishments. For instance, in 1969, institute director Dr. Franklin H. Hop traveled to Toyko, Japan, to make a presentation at the International Conference of Rural Medicine. He was joined by Dr. Keith Long, a toxicology specialist, and by L. W. Knapp, who both also presented papers at the conference.[93] Knapp, who by 1971 was the director of the Institute of Agricultural Medicine, also was "a consultant to the United Nations International Labor Organization and the U.S. Bureau of Labor" and served as the chairman of the National Safety Council's Farm Conference.[94] A brief biographical sketch of Knapp appearing in the *Farm Safety Review* listed his accomplishments and noted that Knapp "has authored numerous bulletins, technical papers, publications and articles, several of which have been published abroad." Despite his prodigious publication record, Knapp maintained his commitment to efforts in Iowa by serving as the president of the Iowa Farm Safety Council in the early 1970s.[95] Such accomplished members were the hallmark of an organization that continually made innovative contributions to farm safety efforts.

THUS, BY THE END OF THE 1940S, the leaders of many organizations were cooperating to provide vigorous safety education in the countryside. The theme of intra-organizational cooperation was demonstrated at the national, state, and even local level. For instance, local businesses, such as implement dealers, banks, and newspapers, often supported the efforts of youth clubs and a variety of other organizations. In 1956, Philip Tichener, an extension information specialist from Minnesota, highlighted the high levels of collaboration that had made a "safety campaign" in Olmsted County, Minnesota, possible. He noted that his state's extension employees had assisted county organizations in a variety of accident prevention activities, including conducting safety exhibitions, distributing educational materials, and visiting individual farmers, all in hopes of reducing farm fatalities and injuries.[96]

The farm safety movement had fully emerged in the first part of the twentieth century, and it continued to evolve and strengthen throughout subsequent decades. The growing influence of agriculture engineers and other experts, the labor crisis of the Second World War, and the widespread recognition of the unacceptably high numbers of accidents contributed to the movement's growth. The efforts of safety advocates were also bolstered by the generous pocketbooks of corporate benefactors who contributed to safety programming. However, despite this dynamic intra-organizational effort, education offered only one possible approach to the difficult problem facing safety advocates and specialists. Thus, experts of many kinds offered diverging solutions and debated the best way to reduce accident rates in the countryside.

Despite the cooperative efforts to raise accident prevention awareness dominating the movement's early years, therefore, a more confrontational tone emerged in the 1960s and 1970s. Farm safety advocates found themselves grappling with a complex array of uncertainties and challenges that had not existed in the 1940s and 1950s. Such developments included the potential impact of the Occupational Safety and Health Administration, the increasing frequency of farm accident litigation, and persistent disagreements about how best to solve the persistent problem of farm hazards. Many experts began calling for greater emphasis on improving product design, for example. Evidence of these new challenges emerges in the records of organizations such as the Iowa Farm Safety Council from the 1970s. In 1969, Lex Hawkins, a lawyer from Des Moines, Iowa, in an address to a meeting of the Iowa Farm Safety Council (IFSC), encouraged farm equipment manufacturers to more fully consider safety in designing their machines. He pointed to the changing legal environment symbolized by the growing influence of consumer advocates such as Ralph Nader. Hawkins indicated he was "making a living off the manufacturers' and distributors' goofs in not protecting the users of machines and equipment."[97] In 1973, the IFSC secretary, Wendall Van Syoc, recorded that "the large committee structure has ceased to function and the work carried on by the IFSC has been more of an individual basis." Such comments suggest that a more fragmented, less cooperative tone had replaced the unified efforts of the movement's early years. Additionally, council members admitted that they were "now in a fast changing scene, the idea of safety has caught on and government influence is being exercised by the USDA, Department of Labor and OSHA." Additionally, the transcripts indicate that some members were troubled that the council had been less active than in the past and thought it might require reorganization to function effectively once more.[98] Such sentiments contrast greatly with the more optimistic and active years of state farm safety committee work.

6

PROFESSIONALIZING FARM SAFETY

A MORE SYNCHRONIZED national farm accident prevention effort emerged during the Second World War with the help of a new generation of farm safety professionals. These experts claimed special authority in reducing agricultural accidents and continually sought to expand their influence. In 1944, Maynard Coe emphasized the farm safety experts' role in "coordinating" and "centralizing" accident reduction efforts. He also praised New York, Wisconsin, and Ohio for becoming the first states to employ full-time farm safety specialists.[1] Eight years later, Coe again pointed out that farm safety experts were needed in all states to fulfill this "coordinating" role and that "all too often, such efforts have been scattered and 'extra-curricular.'"[2] Although Coe's comments adequately describe the profession's initial role, his characterizations are insufficient to describe the specialists' complex experience in influencing and reacting to the intense technological changes in agriculture in the period that followed.

In 1951, A. J. Schwantes, head of the University of Minnesota's agricultural engineering department, accurately described the responsibilities of these new professionals. He stated that they "should act as a catalyst in urging safety activity on the part of other groups on the state, county and local level."[3] Thus, farm safety boosters conceived the role of such specialists as instilling a sense of harmonization in a movement that had previously lacked organization. The profession arose based on the assumption that farmers required expert advice and specialized knowledge to keep themselves and their families safe in a rapidly changing agricultural environment. Schwantes's perspective reflects assumptions within the twentieth century's broader engineering culture. Historian Thomas P. Hughes identified the values of "order, system, and control" as typical engineering ideals of the time. Experts subscribing to such values predictably embraced a new type of professionalism aimed at promoting safer work practices and the larger goal of creating less chaotic workplaces.[4] Unsurprisingly, such systematizing efforts would be complicated by the decentralized nature of Corn Belt agriculture.

The discussion around Glenn I. Prickett's appointment as Minnesota's farm safety specialist clarifies the emerging role of the farm safety expert. In 1949, Paul E. Miller, Minnesota's director of extension, announced Prickett's selection as the state extension farm safety specialist, explaining that Prickett would provide necessary leadership for state extension accident prevention activities and encourage cooperation with numerous organizations throughout Minnesota. Similar qualifications for farm safety experts were part of the job descriptions used in many other states as well.[5] These expectations indicate confidence in the ability of these experts to reduce the frequency of farm accidents through an atmosphere of collective action.

The notion that farmers required outside expertise to navigate the new technological complexities of their professions had been a recurring theme in government and other organizations since at least the early twentieth-century Country Life Movement. Historian David Danbom, in his study of that movement, found that the many urban Americans dominating its reformist coalition believed that rural society had to be improved in response to modernity and that, if reforms were neglected, "the countryside would eventually be populated mainly by ignorant and inefficient people."[6] Many also believed that the nation's farmers were inadequately prepared for the challenges inherent in the tremendous technological changes that were then altering farmwork.

Historian Deborah Fitzgerald notes that the call to modernize only intensified following the First World War, when many stakeholders in agriculture, including "economists, farm managers, employees of agricultural colleges, and particularly farm and home demonstration agents, rural banks and insurance companies, and agricultural businesses," expressed desire for reform.[7] She emphasized that such groups often believed that the agricultural sector represented one of the last parts of the economy limited by disorder and incompetence.[8] Many of these "agents of industrialism," in Fitzgerald's view, believed that agriculture had become so complicated that farmers could no longer navigate it for themselves. She stated that "this simple but powerful idea operated at two levels. First, it corroborated what many urban critics had long expected, that farmers as a group were not smart or hard-working. Second, it reinforced the idea that farm management experts needed to redesign the basic principles of the farm enterprise."[9] The farm safety specialist professionals, confident in their ability to assist farm families in coping with change, were part of the response to these beliefs.

Agricultural engineers of the era echoed these observations, further influencing the development of the farm safety specialist profession, many practitioners of which possessed agricultural engineering backgrounds. In 1937, George R. Boyd, an agricultural engineer employed by the U.S. Department of Agriculture's Research Administration, stated, "The farm factory as well as the shoe factory must be organized, equipped, and coordinated [such] that each one of the separate parts fits into the production program and is coordinated with all of the other parts." Harold E. Pinches, assistant professor of agricultural engineering at Connecticut State College, concurred, stressing that the role of his chosen profession was changing from providing engineering experts for farmers toward "the engineering of agriculture."[10] The increasing tendency among safety specialists was clearly carrying the farm safety movement toward reliance on expert knowledge to address the persistent issue of farm accidents.

By the 1940s, many agricultural engineers had adopted the belief that agriculture required reorganization to prevent accidents. In 1940, E. W. Lehman emphasized this idea while addressing an audience at a national NSC meeting: "You have seen children standing on a harrow or on the back end of a tractor while dad wasn't paying much attention. We need the type of discipline whereby it would not be necessary to caution folks that this or that particular machine should not be ridden on by children."[11] His use of the term "type of discipline" implies that farmers who allow second riders reveal endemic carelessness. Lehman believed such potentially hazardous practices could be reduced through professional education efforts. In 1946, C. L. Hamilton compared the established field of "industrial safety" to the emerging discipline of agricultural safety. He asked readers, "Why can't farm safety develop in the same manner?" He emphasized the great need for improving farm safety and declared it was time for action.[12] W. V. Lambert, head of the University of Nebraska's College of Agriculture, also addressing an NSC audience, stressed his belief that hiring farm safety specialists across the nation would greatly reduce farm fatalities.[13] A. J. Schwantes concurred, indicating that he was pleased by the fact that farm safety specialists had been

Although many safety advocates warned farmers about the dangers of "second riders," many rural residents recalled fond memories of sharing a tractor ride with a parent or grandparent. Persistence of such practices reinforces the point that reducing accidents on family farms involved many challenges absent from manufacturing environments. *F.W. Kent Collection, State Historical Society of Iowa, Iowa City. Reproduced with permission.*

hired in a few states and that many areas were expanding their safety programming.[14]

USDA officials and agricultural equipment manufacturers also called for a more professional approach to accident prevention education. In 1942, a contributor to a USDA farm accident study echoed Lehman's views. The researcher argued that many farmers lacked a proper appreciation of the many potential dangers related to their chosen profession.[15] In 1942, Frank J. Zink, director of the Farm Equipment Institute's research department, expressed similar opinions while addressing attendees at the Nebraska Conference of Home and Safety, held in Lincoln, Nebraska. He stated that the majority of accidents "were due to faulty human behavior" or "carelessness" caused largely by a lack of safety knowledge that could be addressed mostly through a concentrated accident awareness program.[16] Zink also believed that safety instruction was particularly important for farm children, since their parents were usually the only source of safety advice.[17] Zink argued that such teaching had previously been dominated by "volunteers" but these volunteers needed the help of specialists.[18] Given this general call from academics and other authorities for an expertise-based approach to the farm accident problem, farm safety specialists were increasingly recognized as ideally situated to provide the needed leadership and professionalism.

In 1956, a national meeting of 250 farm safety advocates gathered in the nation's capital to discuss the problem. Iowa's Kirk Fox, Illinois farm safety specialist O. L. Hogsett, and several NSC representatives also attended this three-day Farm Safety Clinic. Participants repeatedly expressed their support of the farm safety specialist vocation and acknowledged the multifaceted challenges they faced in their efforts. Participants concluded the deliberations by composing a document outlining eight recommendations to be sent to President Dwight Eisenhower. A *Farm Safety Review* writer indicated that one proposal called for the addition of a farm safety specialist in every state exclusively devoted to agricultural accident prevention. The plan also encouraged "private farm organizations" to hire such experts.[19]

PROFESSIONAL CHALLENGES

Despite such enthusiastic recommendations, specialists' numbers remained small throughout the middle decades of the century. Exact figures are difficult to obtain, but an approximate count can be derived from attendance at Farm Safety Institute meetings and from occasional estimates in brief articles in the *Farm Safety Review* assessing the profession's expansion and status. In 1945, three farm safety specialists, including Randall C. Swanson of Wisconsin, W. E. Stuckey of Ohio, and Katherine Olmstead of New York, attended the NSC's initial Farm Safety Institute meeting in Chicago, Illinois.[20] The following year, five farm safety specialists attended the event, again held at the organization's Chicago headquarters.[21] In 1948, ten states had a full-time farm safety specialist, indicating the profession's slow growth.[22] Such modest gains characterized farm safety specialist expansion throughout the rest of the decade.

During the 1950s enthusiasm for the approach persisted, but expansion of the profession continued at the same sluggish pace. In 1951, A. J. Schwantes indicated that "five states" employed a full-time farm safety specialist as part of their state extension programming. An additional ten were employed either by farm bureau organizations or by state industrial commissions.[23] In 1956, the *Farm Safety Review* reported seventeen full-time farm safety specialists, ten of whom worked in the Midwest. The *Review* did not indicate how many professionals devoted at least some of their working hours to accident reduction efforts.[24] Despite such tepid growth, some safety advocates embraced the farm safety profession as potentially lucrative. In 1959, Harold M. Pontius, a private industry safety consultant addressing a group of adolescents at the NSC youth sectional meeting, stressed that farm safety represented an emerging career choice for young people. That field, like auto, industrial, and public accident prevention, was growing, he said, and individuals with an agricultural background might consider it as a possible occupation.[25]

A 1960 *Farm Safety Review* article updated the numerical picture of the profession's status. Farm safety specialists were now employed by fifteen states, especially in the Corn Belt, including in Nebraska, Minnesota, Iowa, Wisconsin, Illinois, Indiana, Ohio, and Michigan. The Rocky Mountain and Pacific Coast states had at most a single individual employed part-time in farm safety but no full-time safety specialist. California and Wyoming reported having a full-time safety specialist in 1956, but by 1960 California had only a half-time specialist, while Wyoming did not have anyone specifically devoted to farm safety work. States in the Deep South, such as Louisiana, Alabama, Georgia, and the Carolinas, had no farm safety specialists.[26] Thus, the profession continued to represent a rather small occupational group.[27]

In fact, as late as 1988, "fewer than thirty states" had a full-time farm safety specialist.[28] The paltry funding devoted to farm safety throughout the era most likely contributed to such unimpressive numbers. In 1988, Purdue University agricultural engineers Mark A. Purschwitz and William E. Field emphasized this persistent problem in an article entitled "Safety Spending: Is Agriculture Being Shortchanged?" Purschwitz and Field revealed that the USDA and Extension Service had received a little over $1 million for farm safety from the federal government in 1976. However, the authors stressed that even that insignificant amount was "cut to $970,000" ten years later.

They also indicated that "this level of federal agricultural spending funding contrasts sharply with funding for occupational safety in other industries. For example, mine safety efforts received $195.4 million in 1987." The authors also emphasized that neither the USDA nor OSHA currently employed a full-time farm safety professional.[29] Compared to such groups as agricultural engineers (the American Society of Agricultural Engineers, for example, boasted a membership of 3,400 in 1951),[30] farm safety experts were few in number, but they could have a significant impact as the only professionals devoting their entire careers to farm safety.

The farm safety professional was one among many laboring to make farms safer and consequently struggled to establish exclusive authority over the issue among a diverse mix of concerned occupational groups, including engineers, rural educators, and public health officials. Farm safety professionals thus faced unique challenges in solidifying their professional authority. For instance, the chemical safety problem appears to have attracted a particularly eclectic group of experts, including entomologists, toxicologists, medical doctors, veterinarians, and even ophthalmologists,[31] a situation explained by the many types of chemicals in use, the many types of hazards they posed, and the myriad of ways and forms in which they were used by farm families.[32]

Professional boundaries appear to have remained imprecise throughout the era. For instance, Dale Hull, an Iowa State University (ISU) agricultural engineer, devoted much of his time to the farm safety problem, but he did not officially become a farm safety specialist until Norval Wardle's retirement in 1970. Hull first shared the duties with ISU agricultural engineering professor Dr. George E. Ayers. In 1974, Ayers left ISU and Hull became a full-time safety specialist until his own retirement three years later.[33] Farm safety gatherings attracted not only farm safety specialists but also many other experts whose work intersected with the issue. For instance, in 1973, farm safety activists were invited to four regional seminars held at different locations across the country. Speaking to the farm safety movement's professional diversity, the invitation welcomed anyone whose work was affected by the farm safety issue.[34]

This multiplicity of professional backgrounds appears not to have been exclusive to farm safety but rather reflects the general trend in many areas of the broader safety field. In 1957, James D. Hill, director of the Field Service Department at the NSC, stated, "The unvarnished fact is that you can't start in college as a freshman and take a four-year course toward a career in safety. The reason is simple: to be in safety at all, you must be a part of something else, so also you must be good at something else before you can do a job in safety. Every safety man is an 'ex' something else."[35] Hill's contention is logical, since accident prevention in any field requires specific knowledge related to that particular industrial or occupational domain. His observations remained relevant to the farm safety specialist profession throughout the late twentieth century.

Farm safety specialists and their supporters consistently worked to justify the profession's existence by praising their collective accomplishments. In 1968, a *Farm Safety Review* writer indicated that W. E. Stuckey was "convinced that in order to have a really effective farm safety program in a state, there must be a professional worker to originate, stimulate and coordinate activities." The author stressed that the history of the Ohio farm safety movement proved this point.[36] In 1965, Ralph E. Patterson, a federal extension service agricultural engineer, supported the view that the presence of an expert improved the safety situation, indicating that Minnesota's farm families had benefited from farm safety activities. Patterson credited Glenn I. Prickett's efforts with contributing greatly to the reduction of farm mishaps, claiming that between 1949 and 1964, farm accidents had decreased by 50 percent.[37] Such praise undoubtedly contributed to raising the status of farm safety professionals nationally.

NIFS leadership established a committee of professional development as part of the effort to improve their profession's status. Members of the group continually encouraged their associates to publish academic articles, create college-level farm safety classes, and become active in other professional organizations, as spelled out in a 1967 report identifying such activities as crucial to improving their profession's standing. Committee members also encouraged their associates to gain "the recognition" of "other professional groups." These organizations included the American Society of Safety Engineers, the American Society of Agricultural Engineers, and the National Fire Protection Association.[38] NIFS members thus dedicated themselves to furthering their own prestige in the hopes of creating a respected cadre of farm safety specialists.

The farm safety specialist position, however, lacked clear definition even as late as the early 1980s. In 1981, while speaking at the fall National Institute for Farm Safety (NIFS) meeting, Robert A. Aherin, a

University of Minnesota state extension safety specialist, stated, "Agricultural safety and health professionals today, unlike other professional technical areas, are not a very homogeneous group."[39] He stressed that despite the inherent strength provided by such diverse backgrounds, the lack of occupational uniformity did present challenges, especially in terms of strengthening professional identity. Ascribing part of this challenge to the "relatively new status" of the occupation and the lack of standardized schooling in the first decades of farm safety work, Aherin recommended that NIFS members establish a core set of principles and skills that all farm safety practitioners should possess.[40] In the early 1980s, NIFS members presented professional development seminars to educate their members on a core of common knowledge. These "short courses" focused on such crucial topics as understanding the unique aspects of the farm accident problem, the specifics of collecting farm accident data, and many others.[41] In 1981, officials at the University of Iowa Institute of Agricultural Medicine made its facilities available to host such short courses for NIFS members.[42]

PLACES OF EMPLOYMENT

The nation's state universities and extension services employed many leading farm safety experts. This fact should not be surprising since these institutions had long delivered agricultural knowledge to the nation's hinterland. W. E. Stuckey, Glenn I. Prickett, Randal Swanson, O. L. Hogsett, F. R. (Dick) Willsey, and many other leading farm safety specialists were employees of such institutions for the bulk of their careers.[43] Stuckey began his career as an employee for the Ohio Industrial Commission, from 1945 to 1952, before leaving the field for a brief time. In 1956, he returned as Ohio's farm safety specialist, but this time as an Ohio Extension Service employee for twenty years. Prickett led farm safety efforts during his tenure at the Minnesota Extension Service from 1949 to 1964. Swanson remained at his post at the University of Wisconsin from the time he was hired in 1945 until his retirement in 1967. Hogsett served from 1955 to 1983 as an extension safety specialist at the University of Illinois Champaign-Urbana, while Willsey's service as an extension safety specialist at Indiana's Purdue University spanned three decades, beginning in the mid-1940s.[44]

Although farm safety experts most frequently worked in extension services or academia, other possibilities were available, especially in the profession's formative years. In 1946, Walter W. Whitlock was hired as an expert by the Illinois Agricultural Association to help reduce farm hazards.[45] In 1948, James E. Wiles successfully obtained employment as a specialist for Oregon's Industrial Accident Commission.[46] Samuel L. Horst began work as a farm safety professional for Pennsylvania's State Department of Public Instruction.[47] In 1950, the *Farm Safety Review* reported that the state farm bureaus of Iowa and Kentucky had hired two "new state Farm Safety Directors" to promote farm safety awareness.[48] Such miscellaneous titles and employment were typical in the profession's formative years, prior to the solidification of an occupational identity in the 1950s and 1960s.

A MULTIFACETED ROLE

Regardless of the place of their employment, farm safety specialists were faced with a problem of great complexity: As discussed in chapter 2, farm hazards were associated with animals, machines, and chemicals but also with fires, electricity, and many other potential dangers.[49] The experts' contributions to the farm safety movement had to take a variety of forms as well, including investigating the causes of farm injuries and fatalities, collecting accident data, and proposing technical solutions to reduce hazards.[50] Farm safety specialists also helped to fashion and implement diverse state-level educational programming and established and maintained links with other members of the agricultural community. Moreover, their field required staying current both with regulatory developments and with the latest equipment improvements, such as PTO guards and tractor rollover protective devices, which they helped research, test, and promote.[51]

Norval Wardle's career provides a useful case study for understanding the range of experience and knowledge required of a leading farm safety specialist. Wardle was employed at Iowa State University beginning in 1947 and remained there until his retirement in 1970. One of eleven children, his youth had been spent on a farm near Victor, Idaho. Wardle, like many other rural children at the time, began work at an early age, picking potatoes and later planting and harvesting grain with his father.[52] His impressive academic accomplishments and inclination for civic involvement provided him with opportunities despite obvious economic challenges. During high school, he served as student body president and, in 1937, graduated with honors with a degree in agriculture from the University of Idaho.[53]

Wardle taught vocational agriculture in Idaho for eight years and assisted in training Vo-Ag teachers before receiving a master's degree from the University of Idaho in 1946. Wardle later moved to Ames, Iowa, pursuing a doctorate in vocational agricultural education. During his doctoral work, he served as an instructor at the ISU agricultural engineering department and designed two courses. In 1947, he was appointed to the newly created position of agricultural safety engineer at Iowa State University, just before completing his PhD in 1949.[54] Wardle had a deeply personal reason for becoming involved in safety work. An article in *Farm Safety Review* quoted him as saying, "I had seen my father-in-law killed in a farm accident, and was quite familiar with the many dangers and hazards of farming."[55]

Wardle received various public health awards, published hundreds of farm safety articles, and served as the secretary of the Iowa Farm Safety Council during his tenure at ISU. He also served as the first president of the independent National Institute of Farm Safety organization from July 1962 to October 1963, following its official separation from the National Safety Council.[56] Norval Wardle also fulfilled an important role in the effort to increase the professional status of the farm safety field. He was a very active member of the organization's professional improvement committee throughout the 1960s,[57] and he assisted other Iowa Farm Safety Council members to create and implement a variety of programs, actively participating up until his retirement in 1970. Wardle and his contemporaries did much to assist concerted efforts to address an issue of national importance.[58] In 1971, the National Safety Council honored Wardle with the organization's highest individual award, the "Citation for Distinguished Service to Safety," praising his "many years of untiring, dedicated efforts on behalf of farm safety in his state and the nation."[59]

BACKGROUNDS OF FARM SAFETY SPECIALISTS

Norval Wardle's rural Vo-Ag background was shared by many other leading farm specialists. W. E. Stuckey likewise spent his youth on a farm. In the 1930s, he became a Vo-Ag teacher and in 1944 received his master's degree in agricultural education from Ohio University.[60] William D. Lutes of Nebraska had been reared on a farm and served as a Vo-Ag teacher in several Nebraska communities before officially becoming a farm safety specialist in 1954.[61] His successor, Rollin Schnieder, also had agricultural roots; he became a Vo-Ag teacher in Red Cloud, Nebraska, and in 1959 the state's farm safety specialist. Schnieder possessed both BA and MA degrees in Vo-Ag education from the University of Nebraska, and his professional honors included the USDA's Distinguished Service Award and the NSC's Distinguished Service to Safety Award in the 1980s.[62] Glenn I. Prickett had also been a farm boy before working in agricultural education as the University of Minnesota's assistant 4-H club leader.[63] F. R. Willsey, like his peers, had been born in the countryside and served for six years as a vocational agriculture teacher in his home state of Indiana.[64]

The farm safety movement initially emphasized education, so it is not surprising that those with a Vo-Ag background would be drawn into the field. In 1944, W. T. Spanton, the head of the federal government's Agricultural Education Service, a part of the U.S. Department of Education, stated, "I very naturally believe that our nine thousand teachers of vocational agriculture scattered widely throughout the country in as many different rural communities are in a very strategic position to assume and should assume, a major share of the responsibility in the farm safety educational program."[65] Spanton's comments highlight the educational nature of farm safety accident reduction efforts in the movement's early years. Thus, Vo-Ag teachers would serve an important role in a national farm safety movement mainly because they were already providing formal agricultural education. Vo-Ag teachers also moved easily into the new farm safety occupation since they were well trained and positioned to formulate safety educational programs in the country.[66]

Many leading farm safety experts also possessed agricultural engineering backgrounds. This is not surprising, since recognizing and reducing farm hazards often required the skills and technical backgrounds of the agricultural engineer. In 1951, Maynard Coe called attention to the contributions of agricultural engineers in farm safety, stressing their highly significant role in efforts to make farms safer.[67] This trend appears to have remained steady throughout postwar era. In 1985, Glen H. Hetzel, a Virginia farm safety specialist and National

Institute of Farm Safety president, emphasized the importance of his agricultural engineering background in performance of his professional activities, describing his experiences as an "agricultural engineer, teaching courses in the power and machinery areas, as well as various other skill areas" as familiarizing him with issues related to farm safety. Hetzel had also been employed at Virginia Tech from 1967 to 1999.[68]

Many farm safety specialists had a similar background. O. L. Hogsett, a longtime and highly influential Illinois farm safety expert, possessed a BA in agricultural engineering.[69] Fred W. Roth had taught engineering courses at Michigan State College prior to his appointment as an extension safety specialist in 1946.[70] C. E. Stevens, a farm safety specialist employed by the University of Missouri, had previously taught agricultural engineering courses at the University of Delaware.[71] Dale Hull and David Williams both held agricultural engineering PhD degrees. Hull served as Iowa's farm safety specialist from 1970 to 1977, and Williams assumed Hull's responsibilities following his retirement. Herb Plambeck, the author of *Iowa Farm Safety in the Twentieth Century: A History of Contributions of Rural Safety Volunteers*, concluded, "When all is said and done, it must be recognized that agricultural engineers have always been, and probably always will be, key leaders of in farm safety."[72] Plambeck's comments are germane even following the creation of the farm safety specialist as a distinct profession. In 1943, when Randall Swanson became Wisconsin's first farm safety specialist, his office was located in the University of Wisconsin's Department of Agricultural Engineering.[73]

Some farm safety specialists possessed neither an agricultural education nor an engineering background; however, evidence suggests that this occurred most frequently early in the field's development. For instance, Walter W. Whitlock, an early Illinois farm safety specialist, was a U.S. World War II navy veteran with bachelor's degrees in chemistry and education prior to his appointment in 1946;[74] Gordon McCleary, who worked for the University of Illinois, possessed a degree in journalism.[75] In 1949, David G. Steinicke became a Michigan State University farm safety specialist with a master's degree in sociology and doctorate work in sociology and psychology. Although a *Farm Safety Review* writer praised his contributions in developing Michigan's farm safety programs, Steinicke left the post in 1955. His replacement, Richard G. Pfister, had most recently taught agricultural engineering at Michigan State University and had also been a Vo-Ag instructor.[76]

PROFESSIONAL COLLABORATION AND CONTACT WITH FARMERS

Farm safety specialists accomplished much individually, but they nonetheless did not labor in a vacuum: They met and shared insights with their counterparts in other states. The annual Farm Safety Institute meetings provided an ideal setting for such exchanges and for relationship building. In 1945, the NSC and USDA launched the first of these annual gatherings to promote coordination among farm safety experts. The institute attracted a small but growing number of farm safety professionals as well as agricultural engineers, USDA officials, and farm equipment manufacturers. Representatives from other organizations interested in making farming less dangerous also attended.[77] The 1946 meeting attracted specialists from eleven states, including F. R. Willsey, Glenn I. Prickett, Randall Swanson, and W. E. Stuckey. Participants attended a variety of informational sessions over three days, including one on the potential hazards associated with pesticide use and food contamination. Participants also shared lessons gleaned from "outstanding county farm safety projects in Iowa, Indiana, and Ohio" in hopes of designing similar programs in other states.[78]

Other sessions covered topics such as fire hazards, proper use of farm chemicals, and correct electric fence installation. Farm safety experts shared insights regarding ways to improve educational campaigns, and methods of identifying potential hazards. These included ways to increase local media participation as well as the latest practices in carrying out farm accident surveys. Farm manufacturing officials also discussed design changes meant to improve equipment safety. For instance, in 1955, Ford Motor Company's tractor division sent a delegation to lead a panel discussion regarding technical considerations in manufacturing safer tractors.[79] Farm safety specialists thus benefited from the opportunity both to exchange information and to form productive interactions with their peers in other states.

This collaborative spirit expressed itself not only at the annual Farm Institute Meeting but as part of many other professional activities, including consultations when creating state-level educational programming. For instance, in 1956, Norval Wardle sought information from Kansas Farm Bureau Safety Department officials concerning a tractor safety program. Grice E. Sexton, the bureau's safety director, expressed confidence in their safety activities while responding to Wardle's inquiry: "We certainly feel that this accident prevention exhibition is making all the tractor operators who see it more safety conscious." The most vivid aspects of the display involved graphics illustrating how careless driving practices could easily lead to rollovers.[80] In 1961, Stuckey developed a tutorial packet on farm elevator hazards, made available to safety educators throughout the country, that included "a cartoon, color slide series and a model elevator, as well as a script and promotional material."[81] Such fruitful teamwork undoubtedly helped improve programs and information dissemination.

Farm safety experts worked in partnership to produce safety materials for various organizations. In 1961, attendees at the National Institute for Farm Safety produced a pamphlet outlining fundamentals for developing materials on farm equipment safety. The booklet explained basic yet crucial operating instructions helpful for preventing machinery accidents. The NSC's farm division distributed the materials free of charge to 4-H and FFA clubs. Farm safety specialists also supplied more direct support by visiting and advising youth groups. In 1967, Fred Meinke, a Vo-Ag and FFA advisor from Beaver Dam, Wisconsin, acknowledged that much of his local chapter's success was due to the dedicated support of specialists, including Randall Swanson, the state's safety specialist, and Donald Jensen, an emergency preparedness expert. Both individuals were employed at the University of Wisconsin. Meinke had received useful educational materials from them and had also obtained their personal assistance in designing safety programs.[82]

Farm safety specialists' close cooperation with 4-H and FFA signifies how much they needed hundreds of volunteers to carry out their programs. In 1963, Norval Wardle stated that "the farm safety specialist cannot do his job alone. It is impossible for him to reach and work with each member of every farm family. To make any imprint on the tremendous problem which exists, he must be primarily an organizer, a coordinator, a source of reliable facts and information, and an encourager."[83] Wardle went on to emphasize that most program implementation had to be carried out by volunteers. Farm safety experts therefore worked closely with youth organizations in rural education efforts to promote safe tractor driving, fire prevention practices, and many other activities aimed at reducing persistent hazards.[84]

While farm safety specialists primarily offered their expertise and provided educational materials, they also, on occasion, participated directly in program implementation. For instance, in the summer of 1947, the Eastern Iowa Farm Safety Council sponsored a farm safety photo contest in which young people were encouraged to take snapshots of potential farm hazards. Participation was so enthusiastic that the contest attracted 674 submissions. Norval Wardle was one of two judges, assisted by Del Blumenshine, a photographer for the *Cedar Rapids Gazette*. Wardle and Blumenshine selected for first prize a picture of a farmer with his back turned to a bull. Youth who placed in the top five received prizes, and their photos were displayed as part of a farm safety booth featured at the Iowa State Fair. The contests received funding from the Cedar Rapids Chamber of Commerce.[85]

Although farm safety experts undoubtedly played an important administrative role in the farm safety movement, their degree of contact with farmers remains unclear. The experiences of Stuckey shed some light on this, however. As mentioned earlier, Stuckey, like many specialists, had been raised on a farm, so it is reasonable to suppose that Stuckey and his counterparts with a similar background remained personally connected to that community. In 1957, Stuckey wrote an article in the *Farm Safety Review* discussing a recent farm visit. He stated, "Our family was invited to have a Sunday dinner with some friends who live on a farm in western Ohio. After eating too much of their country fried chicken, we leaned back on our chairs to enjoy our last cup of coffee."[86] His own description of his position specified that he spent, "2/3 of the time in the field and 1/3 in his office" at Ohio State University.[87]

Time in the "field" most likely meant meeting with a variety of individuals and groups, such as state extension workers, various safety committees, and the media, however. And although Stuckey indicated that part of his job included providing advice to farm families, this was mostly with the aid of the local extension agent,[88] making it reasonable to assume that his personal exchanges with farmers were infrequent. The multiple, diverse responsibilities of farm safety specialists support this conclusion. Their activities included organizing, researching, and promoting the farm safety cause across an entire state, most likely leaving limited time for direct interactions with farmers.[89] It is also important to consider that farm safety specialists' job duties had become even more complex by the close of the 1960s, probably further lessening opportunities for direct contact. Farm safety specialists by that time were increasingly focused on the potential impact of OSHA regulations and on establishing the identity of their profession. Others, such as Norval Wardle, had also become involved in consulting work.

The Iowa Farm Safety Council minutes confirm the view that farm safety specialists had limited direct communication with farmers. The meeting record from March 1954 included a summary of Norval Wardle's activities during the year. Wardle had spent the bulk of his time assisting in the development of state- and county-level safety events that had attracted more than fifty thousand Iowans. Such happenings included a Safety Day held in Boone, Iowa; 4-H safety awareness campaigns; and harvest safety clinics. Wardle also disclosed his intensive involvement in developing safety educational materials and in conducting several research projects. Although many farmers undoubtedly participated in such events, such tasks most likely limited Wardle's ability to spend time with Iowa's farm families.[90] Wardle, like many safety specialists, was responsible for providing advice and direction for safety activities throughout the entire state.

FARM SAFETY SPECIALISTS' largely educational role was influenced by the more contentious environment that emerged by the late 1960s and 1970s. This was expressed in a number of ways, including diverging opinions regarding the role of agricultural equipment companies in addressing the farm accident problem. Norval Wardle was particularly critical of what he believed to be the failure of tractor companies to install adequate tractor cabs. In 1971, he charged that "there were a few that can withstand a tip, but that is the exception." Wardle also complained that cabs failed to protect farmers from fertilizers and pesticides, stating that most were "poisoning traps" because they did not provide effective air filtration for anhydrous ammonia and other chemicals. Safety experts' willingness to voice such criticisms highlights the increasing contentiousness of the safety movement in the latter decades of the century.[91]

7

SELLING THE SAFETY MESSAGE

SPEAKING AT A National Safety Conference held in Chicago in 1961, Frank Burrows, a traffic safety specialist employed by the Chicago-based Citizens Traffic Safety Board, characterized the need to promote safety to the farm audience as "selling safety." Raising awareness about accident prevention required an imaginative approach, according to Burrows, and he encouraged his listeners to use humor, props, and other techniques to capture attention: "Let me encourage you to use some of these ideas so that you will not only be doing a good deed, selling safety, but will enjoy it more and you will be creating an appetite in people who are not hungry by making them want to do what they ought to do—that is, to do every job the right way."[1] Burrows's allusion to the "right way" echoes the comments of many agricultural experts who believed that many common practices, such as second riders on tractors, were careless. Burrows exuded confidence that an "appetite" for safety could be stimulated where none had existed previously.

Farm safety activists took an approach similar to Burrows's directive to display creativity in educational programs, but they had simultaneously to deal with a host of new issues in farm safety. Jack Burke, a veteran National Safety Council (NSC) farm safety writer, wrote in 1978 that "new accident problems arise and some older ones fade into relative obscurity as the world passes them by or we learn how to deal with them."[2] This changing scene involved the familiar hazards posed by emerging technologies and by the farmers' continued independent, entrepreneurial status. But it also involved some new legal realities as federal laws now restricted certain types of child agricultural labor. As new accident preventative devices, including rollover protective systems (ROPS) and slow-moving vehicle (SMV) emblems, became available, safety experts had to expand farmers' awareness of them and help expand their use. (ROPS hardware includes reinforced safety cabs and structures with various bar configurations to protect tractor operators; SMV emblems, fluorescent orange triangles with retroreflective borders mounted on the backs of farm equipment on roadways, help make those vehicles more visible to other traffic.) Thus, farm safety educators were called upon to be resourceful in delivering programs that addressed these new circumstances.

From the movement's early years, safety advocates had exhibited such inventiveness. For instance, in 1949, the employees of the Auto-Owners Insurance Company, headquartered in Lansing, Michigan, desired to raise awareness of farm dangers at Michigan State College's Annual Farm and Home Week. A *Review* writer described the resulting exhibit, which showcased an electronic device, akin to a pinball machine, featuring pictures depicting various farm tasks such as "Piling Filled Bags," "Using Power Take-Off," and "Leading Bull." Participants in the "Electronic Safety Quiz" had to decide if the person portrayed in the illustration was working in a safe or a hazardous manner.[3] Safety experts actively pursued such innovative approaches, often with support or assistance from private interests, as in this case.

A variety of organizations, including the state farm bureaus, agricultural implement manufacturers, and a variety of local businesses, financially supported safety instruction. Financial contributions ranged from very small donations to relatively substantial monetary gifts. In 1944, J. S. Jones, secretary-treasurer of the Minnesota Farm Bureau Federation, noted that his organization's insurance division had a direct interest in supporting farm educational programming: "I have to admit that we had a selfish motive when we started the safety program. It was started as an issue to our insurance department. We had thousands of claims each year, and being a mutual insurance company where losses averaged over all policy holders, we found ourselves confronted with this situation and losses are paid from premiums—and losses showed us to be in the red for the state at the time to the amount of $500,000."[4] Although Jones's comments reveal some self-seeking motives, such gifts undoubtedly bolstered the movement.

The historical record shows other insurance companies frequently provided both monetary and administrative support for safety efforts. For instance, state farm safety committees often received a portion of their funding from insurance companies. Members of such corporations sometimes served in leadership positions on these committees. L. J. Long Keeny, president of the Grinnell Mutual Reinsurance Company, was a founding member of the Iowa Farm Safety Council and helped provide the impetus for the organization. Harry M. Pontius, the safety director for Farm Bureau Insurance Company, was also an active member during the organization's formative years.[5] The Iowa Farm Safety Council benefited from membership dues paid by other insurance companies and organizations as well, such as the Iowa Farm Mutual Insurance Company and the Iowa Association of Mutual Insurance Associations.[6]

Although farm safety educators exhibited considerable zeal and creativity in their efforts, they also encountered significant obstacles in their attempts to "sell safety" to a rural audience. In 1951, Maynard Coe cited the problem of farmers' "indifference" to safety. He stated that although "it is heartening to note that farm folk in increasing numbers are recognizing the importance of safety, yet there are many who consider 'other matters' of greater importance. Such people practice safety only if they have a little extra time or when these 'other matters' are not pressing them too hard. And such people are the favored candidates for accidents."[7] In 1951, USDA officials sent a questionnaire asking farm safety specialists to list their most frequent work challenges. The experts cited their need for more data to better formulate their safety programs, for improvements in communication with the farm community, and for more effective methods of eliciting positive response to their programs.[8] These problems and requirements appear to have remained constant throughout the period under study.

C. W. Dalbey, an Iowa agricultural educator, indicated that farmers' preoccupation with promptly completing tasks could sometimes override a cautious approach to the work: "I remember many farmers in my adult evening classes who had lost fingers and hands. Often this was due to corn-picker accidents. The attitude seemed to be that the job had to be done now."[9] The perception that farmers were insensitive to safety, however, should be balanced with an understanding of their challenging lifestyle. Safety was only one of many concerns farm families faced in attempting to derive a livelihood from the land. Unlike safety professionals and extension employees, farm families faced a myriad of challenges to their economic survival related to their demanding economic and technological environment. Safety professionals and extension staff, dependent on state and federal funding, were less vulnerable to the vagaries of the agricultural economy.

Individuals who farmed during the era emphasized the often stressful and multifaceted nature of their profession. Don Rimathe, a farmer in Story County, Iowa, emphasized the need to work long hours during harvest time, taking advantage of good weather or running the risk of delays that could harm one's crop.[10] Pat Larkin underscored the profession's fiscal pressures stating, "I never minded hard work, but I hate the stress, you know, handling so much money today; one little mistake [can have profound financial consequences]."[11] Jim Klein remembered that after his father retired from farming, he was finally able to get a full night's sleep.[12] Such comments emphasize that safety was only one of many aspects farmers had to contend with as they struggled to wrest a living from the land. Farmers were independent entrepreneurs, dependent on their own skill, speed, and efficiency. In the face of this daunting gap in perspective between themselves and their audience, safety educators could only respond by harnessing every available means of communication to inundate the midwestern countryside with their message.

PRINTED SAFETY MATERIALS

Safety activists used print as their primary media outlet, distributing tremendous volumes of printed handbooks, pamphlets, and posters produced by both private and governmental agencies.

Although safety information had appeared in earlier farming textbooks and handbooks, the onset of the Second World War gave rise to a proliferation of such materials. The NSC and the USDA fulfilled a major role in this aspect of the movement, but they were just two of many organizations providing safety activists with printed educational resources. Thus, extension employees, the staff of state educational institutions, and even corporations developed their own print materials.[13] Accident prevention boosters depended on such publications to educate themselves and the farm families they work with. As the years progressed, educators could select from hundreds of resources to find those ideally suited to fulfill their constituents' educational requirements. Such diversity, along with the sheer volume of available materials, reflects the increasing influence of experts and government bureaucracy in all aspects of American life. Predictably, as the numbers of rural educators, agricultural engineers, and other experts grew, so did the volume and diversity of their publications.

Even though the NSC had been producing farm safety publications since the early 1940s, the organization's materials continued to develop throughout the era. Farm safety leaders could benefit from an annual dose of information by signing up for NSC's farm safety service for only three dollars. Subscribers received resources that included the *Farm Safety Review*, annual farm accident data summaries, and a register of safety films, as well as the annual NSC congress transactions. Recipients were also sent resources for use in local safety awareness campaigns, including packets of promotional items specifically designed for Farm Safety Week. The NSC continued to produce new safety educational materials to meet the new challenges of agricultural work. In 1973, a *Farm Safety Review* writer announced an NSC-produced handbook, the *Farm and Ranch Safety Guide*, which contained forty pages informing rural Americans about the importance of observing safety practices involving a variety of significant agricultural hazards.[14] In 1974, another *Farm Safety Review* writer celebrated the "31st consecutive observance" of Farm Safety Week and discussed the NSC's continued emphasis on publishing and disseminating farm safety educational materials, including "Program Kits" that had been sent throughout the country to raise the public's awareness of safety issues.[15] NSC staff continually expanded and adapted such materials to accommodate the changing nature of agricultural safety issues, as demonstrated by the numerous and frequent advertisements for educational materials in the *Farm Safety Review* and other NSC publications; cumulatively throughout the era these publications listed hundreds of separate educational documents.[16] In producing such printed materials the NSC fulfilled one of its primary functions. The organization had long generated safety materials for a variety of the nation's economic sectors across a diverse range of industries.

Federal agencies such as the United States Department of Labor (DOL), U.S. Food and Drug Administration (FDA), the Public Health Service, and the Department of the Interior's Fish and Wildlife Division also published farm safety materials. DOL officials appeared to be particularly concerned with the issue of child agricultural labor, as evidenced by their various pamphlets dedicated to the issue.[17] The FDA produced a considerable number of pamphlets directed toward growers, instructing them in how to avoid excess chemical residues. The Public Health Service and the Department of the Interior's Fish and Wildlife Division also provided chemical safety resources, but to a lesser extent than the two other agencies.[18]

As the technological complexity of farming increased, the amount of specialized safety information rapidly ballooned. The diverse number of federal agencies involved in distributing educational material reflected the profound ways in which the increasing technological sophistication of farming affected many groups of people as well as the environment. For instance, the ever-expanding use of agricultural chemicals potentially affected wildlife, urban consumers, and farm families alike. The challenges created by industrial agriculture attracted practitioners of many occupational specialties, who enthusiastically offered their solutions.

The USDA assumed a particularly active role in generating printed safety resources, beginning with the onset of the Second World War. In 1943, Lester A. Schlup, chief of the USDA's Division of Information, indicated that his agency was distributing safety tracts but emphasized that more pamphlets were needed to meet wartime information needs. The government printing office published "[f]ifteen thousand copies of 'Safety for the U.S. Crops Corps'" and only "five hundred thousand each of 'Start Them Right' and 'Going to the Farm Front,'" which resulted in a shortage of materials for distribution.[19] The agency appears to have been especially active in distributing pamphlets aimed at preventing accidental fires and the safe use of agricultural chemicals.[20] However, USDA researchers were also particularly important in disseminating the results of some of the first national farm accident studies. For instance, in 1942, the USDA's Bureau of Agricultural Economics produced a work entitled *The Prevention of Accidents on Farms and in Homes,* which provided a wealth of information regarding the national farm accident picture. Such publications were crucial since they alerted farm safety activists to issues in need of special attention.[21]

State extension officials and university faculty also contributed a significant amount of safety materials addressing almost every conceivable farm hazard. In 1943, John C. Snyder created a series of pamphlets tackling the dangers faced by laborers engaged in the apple and peach harvest. Extension personnel generated materials regarding a host of other issues, including the safe operation of harvesting equipment, electrical hazards, and material-handling equipment.[22] The continued growth in funding, staff, and expert specialties within the USDA and extension undoubtedly influenced the intensified involvement of these agencies in both disseminating safety research and publishing educational materials.

Agricultural journalists also devoted many articles to raising awareness of the issue and demonstrating an ongoing attention to the safety movement long after the wartime emergency had ended. Editors often expressed the view that they were fulfilling an important educational responsibility by featuring farm safety pieces in their magazines. In 1964, Delmer E. Groves, an editor for *Nation's Agriculture,* describing the important place magazines held in the safety movement, noted that "[t]he average farm home receives and reads some two to five magazines."[23] A large number of safety articles appeared in *Wallaces Farmer, Successful Farming,* and a variety of other agricultural publications, attesting to the important part publications played in raising safety consciousness as the movement matured into the 1960s.

NSC officials acknowledged magazine editors who made special contributions to the cause. In 1950, they held an awards ceremony during the annual meeting of the Agricultural Editors' Association in Chicago, at which Sidney J. Williams, an aide to the NSC president, presented eleven agricultural publications with accomplishment certificates. These distinguished magazines included *Wallaces Farmer, Southern Planter,* and *Farm Implement News.*[24] In 1953, a similar banquet recognized the farm press's ongoing commitment to safety. A *Farm Safety Review* writer noted that in that year alone, "a mid-west magazine ran twelve major articles or editorials on farm safety during the year, and devoted its July 19 issue, including a striking cover, to the theme of National Farm Safety Week, entitled 'think safely–act safely.'"[25] The NSC's practice of bestowing such awards highlights the instrumental role the organization played in strengthening the complex mix of public- and private-sector entities participating in the movement.

In 1959, Milon Grinnell, the editor of *Michigan Farmer,* commented on the widespread attention safety had received and suggested additional methods of raising public awareness. He acknowledged that many farmers showed little enthusiasm for the vital subject of safety education, but he believed such efforts could be improved by making the content more specific and engaging. Grinnell thought that general pieces were far less effective than those focused on a single issue, and he provided specific recommendations for improvements. Grinnell also contended that safety writers needed to be more positive in their approach, stating that "[i]nstead of saying children shouldn't drive tractors, it's better to say: children are old enough to drive a tractor when they show mature judgment in completely maintaining the tractor without their fathers having to check on their performance."[26] Grinnell's cautionary advice was prudent since farmers valued their independence and saw themselves as the ultimate authority in managing their operations. Grinnell understood that insensitivity to farmers' independence and autonomy threatened to hamper the effectiveness of safety campaigns.

Robert G. Rupp, an assistant editor for *The Farmer,* a Minneapolis-based publication, indicated that writers who incorporated farmers' accident experiences were particularly effective at capturing readers' attention: "The stories, after the first few had established themselves, climbed up to rank among the highest readership of any editorial content in the magazine." He continued, "Readership surveys repeatedly have shown from 80 percent to 90 percent of all readers, both men and women, read each story." The fact that communities sometimes raised money for injured members of distant farm communities supported the power of such firsthand accounts. Rupp's staff also increased the effectiveness of their safety articles by aligning particular topics with seasonal work schedules. Rupp wrote, "Stories on corn-picker accidents, for example, appear in the fall, just before farmers begin picking corn."[27] Editors like Rupp saw that for safety educators to be effective they would need to be knowledgeable and treat their readers with empathy; they also understood that the quality of their safety articles and pamphlets was as important as their quantity.

Private organizations and business associations in areas closely tied to agriculture also produced printed safety materials, including materials aimed at educating farm families regarding the potential dangers of their products. In 1964, a pamphlet entitled *Chemicals in Agriculture: A Guide for Adult and Youth Programs in Agricultural Chemical Safety,* cooperatively produced by the NSC, the USDA, and the National Agricultural Chemicals Association (NACA), featured an extensive bibliography of chemical-related safety materials. The tract listed five pamphlets and other written materials that NACA members had made available to the public, including *Four Keys to Pesticide Safety, Agricultural Chemicals Safety Code,* and *Pesticides and Safety.*[28] NACA also provided the NSC with many of the written materials NSC required to encourage state and local activists to mount agricultural chemical safety awareness campaigns. A *Farm Safety Review* writer noted that safety educators could also obtain such resources from the Manufacturing Chemists' Association (MCA).[29] These efforts confirm that safety had become an increasingly high-profile issue by the 1960s. Along with the decade's reinvigorated environmental movement, greater attention to consumer safety, and growing influence of environmentalists such as Rachel Carson, farm safety issues attracted the interest not just of insurance companies but of other companies made aware that promoting safety was in their self-interest.

RADIO, FILM, TELEVISION

As such means became available, safety proponents incorporated newer forms of communication in their efforts to disseminate information, including film, radio, and television. These methods were enlisted to alert rural Americans regarding the many potential dangers in their homes and workplaces. Educators in both the private and the public sector produced an extensive collection of safety films. Farm broadcasters often searched for new ways to capture the public's attention to raise awareness of pressing issues, including discussing actual accidents and using sound effects to grab attention. Safety promoters were eager to harness the power of any new tools for disseminating information in their ongoing efforts to inform the public regarding potential threats to their well-being.

Nonprofit organizations and the private sector produced a burgeoning number of farm safety films, often with assistance from academia. During the 1950s, the NSC's farm safety division distributed an ever-increasing number of these short movies. In 1952, the NSC produced a film regarding possible corn-picker dangers entitled *Are You Inviting Corn-Picker Accidents?* The film's narrator discussed the problem's scope, the consequences of careless practices, as well as preventative measures. The NSC staff received support in producing such films from farm safety experts, various land-grant college faculty members, and researchers employed by agricultural implement manufacturers.[30] For-profit entities such as insurance companies also produced films with the aid of university staff. The Iowa division of Blue Cross funded the development of a film entitled *Hands Off,* which used farmers' personal accident experiences to highlight corn-picker hazards. Visual aid technicians at Iowa State College created the film, which also discussed ways farmers could prevent corn-picker mishaps. Although such resources were distributed nationally, it is impossible to ascertain how many farm families viewed them.[31] Nevertheless, it is reasonable to assume that some farm families, at least, saw these films and found the visual information approach helpful. Rex Peterson remembered as a 4-H leader in the 1970s often showing young members safety films on topics such as PTO shaft dangers, obtained from the local farm bureau, and he "hoped that it did some good."[32] The involvement of insurance companies again highlights the consistent cooperation of the public and private sectors in directing and implementing the movement.

Safety advocates also recognized the importance of enlisting radio broadcasters into their cause. Extension researchers had long verified the radio's importance in the lives of rural Americans. NSC staff fulfilled a key role in enlisting radio broadcasters into the effort to raise safety awareness, another affirmation of the organization's crucial role in encouraging cooperation. They encouraged collaboration between different stations, provided broadcasters with audio material, and developed ideas on how best to use radio to improve farm safety.[33] In addition, radio journalists displayed much initiative in raising their listeners' awareness of the importance of living and working safely. In some cases, radio broadcasters even assumed leadership roles in the movement by serving on state farm safety committees or by assuming voluntary NSC positions.

The NSC promoted the involvement of radio broadcasters by providing them with advice but also with opportunities to meet and discuss their efforts. As early as 1943, the NSC provided farm safety announcements, transcripts, and scripts to stations throughout the nation. *Farm Safety Review* writers also provided useful articles, such as a piece entitled "What You Can Do to Help Put Safety on the Air." Such writings helped radio programmers see how they could utilize their medium to spread the farm safety message.[34] The NSC's publicity director and the Farm Radio Safety Committee provided the administrative leadership required to enlist broadcasters in raising awareness. The committee included influential media figures representing NBC, the BLUE Network's *National Farm and Home Hour,* and other important stations. These individuals met periodically to discuss ways to enroll radio journalists in the collective effort to reduce agricultural-related accidents.[35]

By 1946, the NSC was formally recognizing those radio broadcasters who made exceptional contributions in raising the public's awareness. In 1950, Sidney J. Williams, an aide to the NSC president, presided over a ceremony to acknowledge their efforts, just as he had for outstanding contributors in print. A *Farm Safety Review* writer stated, "Fourteen radio stations and three networks also received the National Safety Council's Public Interest Award for exceptional service to farm safety during a luncheon given by the council to members of the National Association of Radio Farm Directors."[36] The following year, NSC members hosted another award ceremony at the Hilton Hotel in Chicago. The dinner celebrated the contributions of the staffs of thirteen radio stations that had marshaled an exceptional effort to increase public safety knowledge. Stations in the Midwest receiving recognition included KFEQ of St. Joseph, Missouri, WHFB of Benton Harbor, Michigan, and WHO of Des Moines, Iowa. These and the other ten broadcasting stations honored had featured a diverse range of safety programming. Such activities included conducting numerous interviews with farm accident sufferers, sponsoring various kinds of safety contests, and placing numerous safety awareness ads on the air. For instance, an NSC staff member noted that KWTO of Springfield, Missouri, featured "fifty special programs and four hundred spot announcements during the year [1948]"[37] Thus, the NSC recognized exemplary individual contributions in hopes of inspiring others to join a voluntary cause.

Broadcasters demonstrated much initiative and sometimes even assumed highly influential roles in the movement. Keith Kirkpatrick, a farm broadcaster with WHO Radio in Des Moines, Iowa, not only regularly included safety spots on his station, he also assumed a leading role in directing safety efforts in his state. He was an important member of the Iowa Safety Council, worked closely with Iowa youth organizations, and even assumed important positions in the NSC's farm division.[38] His station employed creative methods of alerting the public to farm hazards, such as narrations of farm accident victims' individual stories, safety jingles, and contests. For instance, during the winter of 1949, WMT, a radio station in Cedar Rapids, Iowa, offered a prize for a listener who could come up with the best safety device. Howard D. Fountain of Iowa City won by creating a device to prevent stock tank drownings. C. L. Hamilton of the NSC and Norval Wardle of Iowa State University cooperatively judged the radio safety

Children's natural playfulness and inquisitiveness could prove dangerous during both work and play. Youth in the countryside sometimes took unnecessary risks while operating farm equipment on tractors, using elevators, or climbing windmills. *A.M. "Pete" Wettach Collection, State Historical Society of Iowa, Iowa City. Reproduced with permission.*

competition.[39] (Although stock tank deaths may appear unlikely, they represented a very real danger on farms. Farmer Bernard Stodghill recalled a close call he had had as a preschooler while attending a farm sale with his father. He "was chasing ducks or geese around the stock tank" and had followed the birds up a ramp into the tank, breaking through the ice. Fortunately, his dad had been there to save him, but Stodghill believed the situation would have been a lot worse if he had been alone.)[40]

Midwestern radio journalists enthusiastically enlisted youth in the safety effort by sponsoring a variety of activities and competitions encouraging children to develop their own safety ideas. In 1949, Ohio farm radio broadcasters and the state's farm safety council cooperated in creating a slogan and jingle contest. Station managers compiled all the submissions and sent them to a panel of judges, which included state 4-H leaders, home economics instructors, and other safety-conscious individuals. The winners included Dan Barlow of Hudson, Ohio, who came up with the catchphrase, "Safety is a factor that is often sought, but it has to be practiced and can't be bought." Richard Dacklin of Lima, Ohio, triumphed with the maxim, "Leave no pitchforks loose on the farm, but a lot of people don't give a darn. It is easy to get hurt with a pitchfork, too. If you don't put them up, they'll run right through your shoe; may puncture an eye, a leg, or two."[41] Similar contests were held in other states across the nation, including northeastern states such as Connecticut, illustrating the movement's widespread dissemination by the early 1950s.[42]

In 1956, Wallace N. Dudney, a *Farm Safety Review* contributor, provided advice regarding ways to improve radio safety spots. He stressed that announcers delivering such segments should "talk clearly," that the programs should be "personal," and that listeners should be directed to sources to "obtain additional information."[43] Personalizing the safety message was consistent with the recommendation that safety promoters be sensitive to the diverse concerns of farm families.

Rural educators also believed television could effectively disseminate safety information, perhaps even more effectively than other media tools. Miriam J. Kelley, a marketing and consumer information extension expert from Kentucky, wrote, "Television is a natural for presenting any kind of information and offers opportunities for reaching a vast new audience, both rural and urban."[44] Farm safety promoters also hoped to capitalize on this potentially influential technology. A *Farm Safety Review* writer stated, "Imagine actual demonstrations of intricate farm operations brought into every farm home through the magic of television."[45] Such comments were common during the late 1940s and 1950s as the technology's potential began to be revealed.

Campaigners seized on the use of television to educate farm families regarding the dangers present in their environment. In 1952, C. V. Phagan, an Oklahoma extension engineer, and Burnis Arnold, director of the organization's Radio and Television Department, cooperatively produced a farm safety program that aired on station WKY-TV. A *Farm Safety Review* writer indicated that Phagan had incorporated a combination of elements into his demonstration, including "model farm machinery, parts from machines which had been involved in accidents, photographs, and sketches," and that he had addressed a number of issues, including hazards associated with electricity, farm machinery, and livestock. Phagan's documentary also informed viewers about how to identify potential farm dangers and what practical steps to take in eliminating them.[46] In 1957, O. L. Hogsett, an Illinois farm safety specialist, indicated that he had worked cooperatively with his state extension's editorial office to produce farm safety film clips. He assisted in incorporating these short films into a safety education bundle sent by the Illinois extension service to the state's television broadcasters. A *Farm Safety Review* writer pointed out that "all told, twenty-nine stations are receiving television materials."[47]

Safety promoters used television to advance their message throughout the era; however, it appears that they utilized television less vigorously than they did radio. Evidence suggests TV stations did not enlist in the movement to the degree that radio broadcasters had done. In 1956, the NSC gave only three achievement certificates to TV stations, while a vastly larger number of radio stations received the recognition. Although safety educators were probably less successful in utilizing television, they were nevertheless occasionally able to employ its power to communicate. A *Farm Safety Review* writer noted in 1974 that an Indianapolis station, WTTV, observed Farm Safety Week by including the topic in its regular programming. Station managers emphasized the importance of youth safety by featuring the issue on *Chuckwagon Theatre,* a locally popular children's show. The host stated, "Although we have lots of city children watching the program, farm safety rules still apply because these youngsters may be around lawn mowers, garden tractors, and various lawn and garden chemicals. Also, they may be visitors to a farm at some time."[48] Americans' postwar exodus from the countryside to cities and towns was making agricultural concerns less relevant to the increasingly urbanized television audience.

Other social developments and circumstances help explain why television may have been less heavily invested than radio in furthering the farm safety movement. First, television was still a novelty during the Second World War when farm safety concerns emerged as an organized movement, unlike radio, which was already long established as a means of communication. The first practical and relatively affordable television sets only came into existence after the war. For instance, RCA's model 630-TS had sold 10,000 units by the end of 1946, whereas a year later U.S. television makers had sold 250,000 sets in the country. Second, the majority of the nation's early television audience resided on the East Coast, with 20 percent living in the New York metropolitan area as late as 1950.[49] Historian Gary E. Edgerton stated that throughout much of the early 1950s, "urban areas were well saturated with TV, but the more isolated hinterlands had few sets."[50] Safety advocates may have been dissuaded by this imbalance from using television for their message. Finally, it is also reasonable to assume that television programmers, recognizing the ever-declining rural population, catered to an urban audience.

DIRECT INSTRUCTION AND DEMONSTRATIONS

Although farm safety activists' use of print, film, radio, and TV was a crucial element in their accident reduction efforts, direct instruction continued to play an important part in their educational work. Demonstrations of agricultural machinery hazards were some of the more popular events. Farmers sometimes saw such presentations during larger agricultural contests, such as the popular corn-picker matches and tractor pulls. Advocates also developed occasions exclusively devoted to disseminating safety messages, such as fire-fighting simulations and tractor-tipping displays. Such exhibitions served to highlight dangerous habits as well as safe-operation practices, and they tended to attract large crowds due to their dramatic and competitive nature.

Safety proponents' use of the in-person approach addressed a variety of farm dangers, including the rural fire problem. In 1949, Chuck Forbes, a KASI farm broadcaster, became concerned with the rural fire threat as a result of several incidents near his hometown of Ames, Iowa. He quickly discovered that members of other organizations and institutions, such as the Iowa Retail Farm Equipment Association, the Ames Fire Department, and Iowa State College, were also concerned about the issue. Forbes, with the aid of individuals from these organizations, worked to create the First Annual KASI Farm Safety and Rural Fire Prevention Field Day, held "on a farm south of Iowa State College at Ames, Iowa." Event organizers were pleased at the large numbers of local farmers who attended and watched the activities, which included observing firemen putting out an actual fire, implement dealers demonstrating the importance of using protective guards, and exhibits advertising other safety equipment. Activity leaders also installed a "five-thousand gallon concrete tank" at the location to show the importance of having an adequate water supply to fight fires.[51]

The greatest enthusiasm among safety experts and the farm community, however, emerged in response to tractor-tipping demonstrations. In 1953, the University of Nebraska's agricultural engineers conducted one of the first of these events at the state's Tractor Field Day. The display primarily involved showing farmers the various kinds of operator decisions that could result in turning over a tractor. Two dummies helped with the demonstration. The first, named Jughead, represented a reckless farmer. Daniel Kitchen, a University of Nebraska at Lincoln agricultural engineer, explained: "Jug-head represented a common type of operator, always hurrying, never stopping to think, and who fails to recognize danger until it is too late." Another dummy, named Bozo, signified the dangers associated with "showing off" while operating with a second rider. Kitchen stressed the possible tragic consequences of such behavior: "Bozo, with little brother in his lap, thrilled him with tractor capers, zig-zagged, turned corners at eight miles per hour . . . the tractor rolled over, fatally crushing Bozo and his little brother."[52] Although the names of the demonstration dummies were humorous, they encountered real situations. Farmers often worked fast, tractors often carried second riders as farmers struggled to meet the overwhelming demands of the harvest, and children worked alongside their parents as they learned the farming profession. Bernard Stodghill clearly and powerfully expressed this reality when reflecting on his years of working with his dad on the farm: "Wherever he was at, I was there."[53]

Safety activists hosted another tractor rollover affair in Boone, Iowa, shortly after the Nebraska event. C. L. Hamilton, an NSC agricultural engineer, stated, "People who pass up articles and lectures on safety stood up in the hot sun for over an hour to catch every point demonstrated. The exhibition's popularity was verified by a flood of requests for repeat performances." Hamilton indicated that more than twenty thousand farmers had viewed the Iowa and Nebraska demonstrations. These dramatic displays proved to be much more effective than the earlier use of small tractor models in instructing farmers about the dangers of careless tractor driving practices.[54] Kitchen agreed with Hamilton's sentiments regarding the usefulness of such realistic demonstrations. He stated that "[t]his full scale visual education was highly effective. Numerous local groups requested repeat performances. With equipment scheduled ahead and demonstrations promised throughout Nebraska, many requests cannot be met."[55] Both safety experts correctly assessed the educational value of such events since they appeared to better capture the public's attention than potentially less vivid educational approaches such as print and radio. The popularity of such dynamic accident prevention exhibitions was enhanced by the inclusion of simple and practical steps to avoid catastrophic rollovers. Safety literature and visual materials were not only less immediate in appeal, but they also in some cases offered more elaborate advice that could be impractical to apply.

In 1954, a *Farm Safety Review* writer described the success that safety advocates outside the Midwest were having with tractor-tipping events, noting that such events were being held in rural areas throughout the nation. New Hampshire safety educators, for example, had great success with their rollover exhibitions. The state's farm families viewed their first tractor-tipping event at the Coos County Farmers' Field Day, and a *Review* writer vividly recounted that "over five hundred attentive spectators saw the roaring tractor tip and thud on the turf." Local implement dealers provided many of the tractors and much of the other equipment required for the event, as well as much of the necessary financial support.[56] Local businesses, like their national counterparts, recognized the public relations value of supporting such efforts, which encouraged the best possible operator practices.

Instructors who lacked the resources required for elaborate full-scale tipping demonstrations could choose simpler options, such as tractor safety kits. Safety advocates continued to purchase small tractor model kits well into the 1960s. Specialists could obtain such teaching aids from companies that manufactured realistic miniature tractor replicas. In 1958, the advertisers employed by MO-RE, Inc., of Bonner Springs, Kansas, promoted such scaled-down tractor models. One company's advertising pamphlet described the tractor's quality: "The tractor is electrically powered with a special 12-volt direct current motor. Its gear box is designed with the precision of a fine watch." The model tractor came with a variety of other useful teaching materials, including a guidebook, a small ramp to demonstrate points of instability, and charts to assist in conveying the information. The booklet advertised the product's many benefits, stressing its usefulness in teaching safety in school classrooms, at county fairs, or in farm organization meetings.[57] Although educators often commented on the superiority of large-scale tractor demonstrations, specialists continued to use these models due to their affordability and practicality.[58] This flexibility and ingenuity in delivering tractor-driving instruction reconfirms the adaptability safety advocates displayed in promoting their message.

Safety activists also used mechanical corn-picking and tractor-pulling competitions to reach rural audiences. The judges of such contests evaluated the participant's attention to safety in assessing performance. Chuck Worcester, a Cedar Rapids, Iowa, radio broadcaster, described referees investigating contestants' corn pickers to determine whether they were operating the machines properly. The judges deducted points from a farmer's total score for not working safely. Worcester stated, "The safety score card used in judging the contests is divided into three sections: equipment on the tractor and picker, safety features, and operation."[59] Safety specialists also assisted event organizers in developing appropriate operating criteria.[60] Operators who displayed careless practices were sometimes even disqualified. In 1952, Harold J. Schmitz, a Missouri radio broadcaster, recounted how seriously some of the contest's officials regarded safety and how the judges' actions left a strong impression on the audience. He recalled that at one event, "Levi Caraway, from Jamesport, Mo., was conspicuously disqualified before a crowd of over five thousand people because he got off the tractor and left the corn-picker running."[61] In 1953, O. I. Berg, a University of Wisconsin agricultural engineer, indicated that safety had been included in his state's tractor-pulling contests by integrating a penalty system. Thus, safety boosters incorporated the accident prevention message into the corn-picking and tractor-pulling competitive farm events rural Americans had long enjoyed.[62]

REGULATORY CHANGES AND DESIGN IMPROVEMENTS

Although safety instructors' creativity did much to shape safety programming, other factors also influenced the nature of accident-prevention education. An altered legal environment and greater recognition of the dangers associated with specific technologies were also important determinants, and these came strongly into play in the 1970s. Evolving legal realities included restrictions on child labor in agriculture, which contributed to the creation of more standardized forms of safety training, particularly for tractor driving. Recognition of the dangers associated with increasingly diverse types of agricultural chemicals contributed to more formalized types of safety education regarding these potentially dangerous substances. Activists also diligently promoted new safety devices, such as ROPS and SMVs. Even though safety instruction evolved in response to these new regulatory and technological realities, education remained a crucial tool in efforts to reduce accidents.

In responding to the altered regulatory environment, vocational agricultural instructors created youth-oriented safety classes on how to meet the requirements of the Hazardous-Occupations Order in Agriculture. This legislation represented an amendment to the Fair Labor Standards Act, barring youth under the age of sixteen from certain kinds of agricultural work on farms not owned by their families. Students under sixteen who completed training, however, received certificates that "exempted" them from some of the restrictions.[63] Such courses were part of the response by safety educators to the new regulatory demands. Originally, farm safety educators had adopted a rather general approach to raising awareness about farm hazards; however, the enactment of the Hazardous-Occupations Order in Agriculture contributed to an increased formality and specificity in such programs. Despite the legal changes, however, family farmers in the Corn Belt still relied almost exclusively on their own families to operate their farms. Thus, most farms were likely little affected by such legislation, demonstrating again that farming in the Midwest proved much more difficult to regulate than other economic sectors.

In 1970, Michigan State agricultural educators responded to farmers' labor concerns by developing new safety courses. They stated that "[t]he intent of the order was to protect the well-being of youth.

It presented problems to farm operators who had traditionally hired youth to cultivate, drive tractors, and perform other tasks now listed as hazardous." They also stressed that new child protective legislation increased the pressure farm owners placed on vocational agriculture (Vo-Ag) teachers to provide more farm safety education opportunities. Such observations reveal that although these classes did represent a positive step in providing youth safety training, the impetus for the courses essentially arose from growers' concerns about meeting their labor needs. As noted in chapter 5, earlier circumstances had linked labor concerns with an interest in safety training: Fear of labor shortages also precipitated improved safety awareness during the Second World War.[64]

Youth safety classes provided students with practical safety training, especially regarding safe tractor operation. Vo-Ag instructors, owners of agricultural implement businesses, and extension specialists cooperatively produced and supported such instruction. Charles E. Wilson, a U.S. Bureau of Labor official, indicated that these local safety classes were met with enthusiasm from the agricultural community. Wilson stated that private businesses had been particularly supportive, and he stressed that many farm equipment dealerships had donated the use of their facilities and provided equipment to aid in implementing these courses,[65] another example of the important support provided by the private sector to promote farm safety efforts.

Although the training was originally developed for youths working on farmsteads not owned by their families, young people attended with different motivations as well. In 1970, Wilson indicated that a substantial number of young persons taking the safety classes worked on their own family farms: "In a recent survey of the students in one area, it was found that less than 50 percent planned to work off the farm home during the summer." Wilson also noted that many students who met legal age requirements lacked agricultural equipment training and wished to acquire such skills; for example, one sixteen-year-old Iowa boy possessed little tractor-driving experience but wanted to learn how to safely operate one before beginning a farm job.[66] Thus, such tractor-driving courses probably had a positive effect on improving the skills of agricultural workers beyond the target audience of hired agricultural laborers.

During the late 1970s, Jim Klein enrolled in a tractor safety program offered by the Iowa State University Extension office in Sibley,

Iowa. Several of his brothers also took the class, which demonstrated how seriously his father regarded safety issues. Jim participated in the course for a few hours every morning over a few weeks and received a certificate after completing the course. Klein recalled watching a film about PTO hazards that showed a pair of shredded trousers from an accident as a warning to viewers. He said that although some students took a casual attitude toward the safety class, such footage had a sobering effect. Sadly, Klein was reminded of the dangers of a PTO on another occasion when a friend was killed in a PTO accident. His classmate's hooded sweatshirt got caught in the spinning shaft while the boy loaded hay bales onto the elevator. The need for safety classes during the era was clear, as was the contribution of safety films to conveying the gravity of the farm safety issue.[67]

By the late 1960s, agricultural chemical safety education had also become more official and standardized. In 1969, Dr. L. C. Gibbs, an administrator for the Federal Extension Services Agricultural Chemicals Program, discussed the success of his organization's "applicator schools." Instructors at these seminars sought to convey general safety principles to those charged with applying chemicals. The seminars also provided advice on how to avoid harming bystanders and polluting the environment and on how to maintain application equipment. Gibbs described attendees as "aerial and ground applicators, pest control operators, representatives of chemical companies and associations, State and Federal government agencies, and university research and extension workers." He noted, "About five thousand persons have attended the eighteen schools that have been held in the past two years in the northwestern, southern, and western regions of the country." The development of such schools revealed that government officials acknowledged that the proliferation of agricultural chemicals required an increased emphasis upon safety.[68] Such training also demonstrated that, although farming was still relatively unregulated, agricultural chemicals were attracting greater governmental attention. Unfortunately, farm families had been using ever-increasing numbers of such substances since the early 1940s, long before the developments of such formalized chemical training.

The emergence of ROPS influenced farm safety education as well. Even though workers in the lumber industry had been using roll bars and protective canopies since the late 1950s, the farming community adopted them much more slowly. This difference was likely attributable, at least in part, to timber companies' greater financial resources as compared to individual farms. Farmers were independent entrepreneurs and could choose whether to purchase tractors with ROPS; timber companies operated in a corporate environment that placed increasing emphasis on safety. Nevertheless, by the early 1970s, *Farm Safety Review* writers were directing safety activists to encourage farmers to purchase tractors fitted with ROPS or to install such protection on older models. Efforts included writing articles for the farm press and showing informational films to rural audiences regarding the benefits of operating ROPS-fitted tractors. By 1973, both International Harvester and John Deere had produced films highlighting the advantages of such devices. NSC officials recommended that safety activists hold tractor accident demonstrations highlighting how the new technology prevented serious injury and death.[69]

For instance, in 1970, a tractor rollover demonstration held at Nebraska's Tractor Power and Safety Day featured a tractor fitted with a protective cab. A *Des Moines Register* reporter wrote that the tractor's interior remained structurally intact despite undergoing significant stress. The journalist commented on the pictures featured in the article stating, "The particular cab shown here survived not only this overturn, but also a crushing resistance test carried out with the aid of a heavy pendulum, and another test involving three backward flips." These demonstrations encouraged farmers to install ROPS or to purchase tractors that already possessed them.[70] By the early 1970s, the NSC's staff vigorously promoted ROPS use by instigating their "Tractor Overturn Prevention and Protection (TOPP) program." The NSC effort included the distribution of educational materials in addition to cooperating with other organizations in hopes of increasing farmers' awareness of the life-saving advantages of ROPS.[71]

ASSESSING SAFETY PROGRAM EFFECTIVENESS

Farm safety leaders not only devoted themselves to continually refashioning their programs to suit new safety equipment, they also assessed their programs' effectiveness, generally giving them mixed reviews for levels of participation. While organizers often boasted of the numbers attending a tractor-tipping demonstration or similar events, some safety activists expressed general frustration at the public's lack of interest in other programming. In 1970, Page Bellinger, chairman of the American Society of Mechanical Engineers Safety Committee, admitted disappointment regarding with the amount of attention the issue had attracted. He stated, "You've heard it many times: 'farmers won't buy safety,' but why not? They buy other things they consider essential for a productive effort. Why not safety?" He also stated, "Early efforts did not produce an avalanche of response, so safety received a back-page treatment, except by a handful of dedicated persons." His comments mirrored the assessments of other safety proponents who had earlier concluded that those in the agricultural community enthusiastically embraced technological improvements related to production but lacked equal zeal for working cautiously.[72] Despite such assessments, it is possible that farmers' lack of enthusiasm for such education did not necessarily mean a lack of interest in safety. Farmers had limited time to devote to safety programming, especially during a period of tremendous technological change and of financial pressures that threatened their economic survival.

Some evidence also contradicts these pessimistic appraisals of farm families' lukewarm response to the safety programming. As early as 1953, George Small, a farmer from Clinton County, Iowa, mounted his own safety campaign. Small had been forced to leave farming after his arm was amputated in a corn-picker accident, and he sought to raise awareness of the need to be cautious around corn pickers by creating a "countywide card campaign." He, along with some local farm youth, distributed signs reading "Will You Be Next to Swap Your Hand for a Hook?" These posters were "distributed to all farmers through school children and 4-H clubs to hang on corn pickers where operators could see them constantly." A *Farm Safety Review* writer stated, "As a result, no corn-picker accidents were recorded in the county during the harvest season. In the previous harvests, there were three arm and four hand or finger amputations from corn-picker accidents in Clinton County." Farm safety programming often contained similar examples of community spirit that then translated into enthusiastic participation.[73]

Regardless of farmers' interest in the subject, safety advocates argued that educational programs had a positive effect in reducing accident rates. In 1965, Ralph E. Patterson, a Federal Extension Service agricultural engineer, stated, "It has been proven many times under many circumstances that where there is an active safety program, there the accident rate is reduced." He cited a reduction in accidents in Minnesota and Georgia, both of which had active rural safety programs. Patterson indicated that Minnesota's farm families had benefited from a farm safety program directed by the state's farm safety specialist, Glenn Prickett, noting that between 1949 and 1964 farm accidents had been reduced by 50 percent.[74]

Despite Patterson's comments, ascertaining the ultimate effectiveness of educational efforts presented considerable difficulties. It is important to remember that statistical data on farm safety was often incomplete and that accidents might have occurred less frequently for reasons other than an education program. For example, safety writers acknowledged that some of the older types of machinery appeared to be potentially more dangerous than newer types. When the combine replaced the corn picker in the 1960s, accident rates were likely reduced. Use by safety activists of numbers of educational event attendees as a singular measure of success would serve as only a superficial indication of a program's effectiveness in preventing accidents. Despite such criticisms, however, safety activists did identify ways to measure the effect of their efforts.[75]

Farm Safety Review writers and attendees at the annual farm safety division meeting continually offered one another advice on better ways to both disseminate the accident prevention message and improve the quality of programming. Writers and speakers emphasized such techniques as personalizing the information, selecting the proper words to communicate with the public, and continually seeking ways to capture audience interest. In 1967, Jack Burke noted that detailed planning should precede every safety campaign or program and stressed that preparation should include obtaining a detailed understanding of the problem, eliciting the assistance of the local community, and creating a sound strategy to carry out the program. He also suggested that activists should thoroughly assess their performance after concluding a safety campaign.[76]

As the safety education movement progressed, rural educators gained additional insights regarding the most effective ways to reach farm families. A particularly fruitful realization was that people were more likely to listen to and internalize information that was presented by people from their own communities. In 1968, Robert E. Kowalski, an Iowa State University extension official, reiterated this point, noting that the public was more likely to absorb messages communicated by members of their own community than by outside specialists. Kowalski revealed that his fellow extension colleagues had recently used this finding in their effort to raise awareness regarding the dangers of placing "household chemical" containers near food products. They had success with placing posters alerting women of the issue in local grocery stores, because, as Kowalski noted, "Homemakers shop in supermarkets which give them the most for their money, and in which they can place their trust concerning quality of products. They eventually establish friendly relationships with their grocer, perhaps on a first-name basis."

Thus, safety proponents became increasingly perceptive in their approaches as the movement matured.[77] Such sensitivity was required, since the majority of Corn Belt farmers were independent entrepreneurs who were sometimes skeptical of outsiders. This understanding was not entirely new, since the perception that farmers, regardless of where they lived, were a particularly independent group had long been held by many observers. For instance, in 1915, Dr. John S. Hitchcock, a Connecticut health officer for the State Department of Health, expressed the sentiment in an article entitled "The Health of the Farmer," which appeared in the *American Journal of Public Health.* He wrote, "He is the King on his place and he keenly resents suggestions from outsiders, particularly if unsolicited, in regard to matters touching his own daily living." Hitchock's analysis is at least partially supported by some farmers' views of outside expertise.[78]

Comments from James Wilson also illustrate that many farmers valued their independence and confidently faced the many challenges of their work. Wilson recalled that when an extension agent in the 1930s provided his father and neighbors with hog farrowing advice, the farmers only accepted the advice they believed was useful and relevant. Wilson recalled his father saying, "We are pretty dang independent and if some city slicker comes out and tells how to do something, we're just going walk off and leave him."[79] Such comments illustrate the sensitivity safety experts needed to craft effective educational programs. Farmer Bernard Stodghill also noted that farmers' life experiences greatly influenced their understanding of and willingness to accept expert advice.[80] In 1966, Larry Petersen, a Plymouth County, Iowa, farmer who had recently been chosen by the local chapter of the Jaycees as the area's "Outstanding Young Farmer of the Year," expressed similar sentiments. Reflecting on his work, Petersen stated,

"I try to keep up with what's new in farming, which means reading magazines and so on. I don't always follow their advice but I do pick up hints here and there."[81] Such comments confirm farmers' autonomous natures and the need for safety advocates to acknowledge such perspectives when implementing educational programs.

Many farmers also believed that agricultural knowledge was passed down through generations, principally by working at farming from an early age. Farm safety advocates had to recognize this belief when crafting youth safety programs. James Wilson discussed this cultural value when discussing his practice of allowing his young son to operate equipment: "If you don't let them do it when you're around, then when you're not around, they'll get hurt."[82] He went on to say, "As soon as they were old enough to do something, I had them doing things," and that once when his brother had criticized him for letting his boy drive a tractor at a young age,[83] Wilson had responded by saying, "If he doesn't know now, he's never going to [know how to drive it]."[84] Most farmers had gained their knowledge of farming, including safety practices, from their parents. Although the farmers interviewed for this project varied in the ages at which they assumed farm responsibilities, they all expressed a similar opinion about how such knowledge was best acquired. This factor meant that farm safety educators faced challenges distinct from those facing accident prevention educators in other economic sectors.

FARM SAFETY EDUCATORS were generally responsive to the changing realities of the sources and causes of farm accidents, and they attempted to raise the rural public's awareness using various communication technologies and teaching techniques. In the decades immediately following the Second World War, activists produced increasingly diversified safety education materials and accident prevention programs. Nonetheless, they faced a number of obstacles in disseminating their message into the countryside, including the constant demands of addressing changing conditions and of overcoming public indifference. Safety leaders also appear to have devoted insufficient resources to assessing the effectiveness of their programs. In all fairness, however, attempts to determine the connection between their labors and a reduction in accidents was problematic. Nevertheless, safety specialists learned that safety programs that encouraged participation or elicited excitement from the audience were superior to less dynamic educational approaches. For instance, youth-centered safety programs were particularly effective at eliciting community involvement and injecting excitement into programs. By the 1960s, approaches other than education had begun to gain prominence in accident reduction. For example, safety activists increasingly emphasized equipment design improvements, such as using ROPS to prevent tractor rollover fatalities. Agricultural safety education, however, continued to represent a crucial component in efforts to reduce the numbers of accidents in the relatively unregulated environment of Corn Belt farming.

8

AN IDEAL FIT: 4-H AND FFA

IN 1949, KENNETH H. ANDERSON, a member of the National Committee for Boys and Girls Club Work, stated, "For the time, money, and effort expended, I believe that more can be accomplished with youth in the safety field than adults. I say this because youths are open-minded and receptive. They are joiners and like to participate in organized efforts. They usually have more time than adults and their persuasive powers with parents and neighbors are great."[1] In 1956, Maynard Coe, director of the NSC's farm division, recognized the enthusiastic participation of rural youth in safety activities: "They are truly missionary in their zeal to make the families in their home communities safer at work and play." Safety leaders also recognized the growing involvement of 4-H and Future Farmers of America (FFA) organizations not only in the Midwest but across the entire nation. For instance, 130,000 Texas 4-H youth participated in accident prevention activities in 1956. Willie L. Ulich, the vice president of the state's Farm and Ranch Safety Committee, stated, "Hazard Hunts, Safety Meetings, and workshops are common to practically all our community clubs."[2] Youth groups in the Corn Belt, including 4-H and FFA, displayed similar levels of activism throughout the era.

Corn Belt families' deep-rooted connections to 4-H could powerfully affect those families' lives over generations, as illustrated by the experiences of Charles and Rose Scott, a farming couple from Monroe County, Iowa. The Scotts established individual 4-H clubs, served as club leaders, and watched both their children and their grandchildren benefit from 4-H membership. Rose served as a 4-H club leader for more than thirty years, confirming her positive view of club work. In her view, youth not only gained knowledge about the rural lifestyle but honed such lifelong traits as confidence, dependability, and pride in their work. Devotion and confidence in the 4-H organization such as the Scotts displayed allowed local clubs to provide an indispensable conduit for accident prevention education.[3] Such community loyalty meant that safety programming attracted more participation than if outsiders had initiated it.

FFA chapters also had a fixed place within local communities, and their close connection to high school vocational agriculture (Vo-Ag) education programs made them crucial to the dissemination of agricultural knowledge to the next generation.[4] Vo-Ag instructors provided advisory leadership for local FFA chapters, while the chapter members assumed lead roles in choosing programming, conducting meetings, and participating in service projects. Members were encouraged to develop activities suited to the local area, allowing FFA programming to differ geographically. Projects included raising livestock, attending guest lectures, and cooperating with other civic organizations.[5]

Individual achievement was rewarded, but so was cultivating a commitment to teamwork and social responsibility. These FFA values were shared with the 4-H organization, making both groups ideal for promoting agricultural safety. *The Official Manual for the Future Farmers of America*, published in 1972, listed "four grades or degrees of active membership: 'Green Hand,' 'Chapter Farmer,' 'State Farmer,' and 'American Farmer.'"[6] To move through these ranks members were required to display notable and expanding degrees of leadership, organization, and personal responsibility. FFA required a significant time commitment, since the organization's official manual stressed the importance of having "at least one planned meeting held each month when sufficient time is available to take care of all chapter business."[7] Some FFA chapters had more than one monthly meeting, however, and many additional planned events, such as field trips, state meetings, and community service activities.[8] Members were encouraged to learn and apply the latest farm practices in their livestock, crop, and horticulture practices. Such training coincided with the goals of agricultural engineers, safety specialists, and extension staff, all of whom desired community youth to accept the latest agricultural knowledge so it could be circulated and adopted by the nation's farmers.[9]

This multiplicity of benefits helps explain why rural youth organizations were involved in the formative years of accident reduction efforts. In 1943, the National Committee on Boys and Girls Club Work cooperated with the Federal Extension Service by enacting the Farm and Home Safety Campaign. The national program was patterned after the Minnesota and Kansas 4-H safety programs, which had been established in the mid-1930s.[10] The nationwide effort demonstrated that the national 4-H leadership was committed to enlisting their organization into the cause of safety, an interest verified by the participation in safety programming of youths from hundreds of local chapters. In 1946, a *Farm Safety Review* writer reflected on the rapid expansion of youth-oriented efforts, stating, "Leading the nation in practical farm accident prevention work for another year, thirty-five thousand 4-H club boys and girls have completed their 1945 safety activities." The writer went on to congratulate rural youths' "immeasurable" contribution to the effort to reduce farm accidents.[11] Another writer indicated that the club leaders who participated in the program hoped to "draw members' attention to common accident hazards on farm and in homes and encourage them to participate in programs and promote safety as a 4-H principle."[12]

Safety advocates acknowledged both the advantages of aiming accident prevention programming toward young people and the eagerness with which those young people participated in them. In 1940, T. A. Erickson, an official for the Minnesota Extension Service, articulated this sentiment at the NSC farm division's annual meeting: "They are still plastic, they are still receptive to ideas, they become enthusiastic about things, and they can talk to dad and mother and put these ideas across to them where a grown-up person wouldn't be able to do it."[13] Erickson's comments would be echoed by Robert Kowalski's views in the 1960s that people were most likely to be receptive to educational messages delivered by people they knew rather than by outsiders. Erickson's statements also indicate that adults were not always receptive to the safety message, but youth were less set in their ways. Although Erickson and Kowalski did not make the point, changing the behavior of youth would have longer-term effects than changing the safety attitudes of older farmers.

Youth groups shared some of the same challenges that other organizations faced in raising safety awareness, such as convincing local citizens of the need to address the problem and gaining their participation in safety activities. An Ohio youth delegate to the National Safety Council (NSC) farm safety sectional meeting also indicated that young people in that state needed more reliable farm accident statistics and wider cooperation among other organizations to improve Ohio's program.[14] Nevertheless, local youth groups possessed some advantages, including their more direct understanding of safety issues in their own neighborhoods and their access to local resources.

Safety advocates often expressed the view that when safety messages reached a community's youth, it assisted in raising accident prevention awareness in adults.[15]

The 4-H organization emphasized achievement and recognized exemplary accomplishments, encouraging individual initiative, and the national 4-H Farm and Home Safety Campaign directors recognized youth who devoted special efforts to the cause. In 1950, for example, the Jackson County, Arkansas, 4-H chapter won a special commendation. Club members had divided themselves into "four committees," each dedicated to a specific farm danger: fire, home, highway, and farm hazards. Helen Baker, a particularly active member, gave farm and home safety demonstrations to a combined audience of two thousand people. Both individuals and club chapters with exceptional achievements received prizes, including free trips to the National 4-H Congress, cash awards, and a plaque. Participation in the National 4-H Farm and Home Safety Campaign expanded from one hundred thousand in 1943 to approximately four hundred thousand seven years later,[16] and participation remained high and consistent throughout the 1950s and 1960s. In 1968, a *Farm Safety Review* writer reported that 516,600 youth had participated in the 4-H safety program.[17]

FFA leaders also enacted a national safety program that attracted an enthusiastic response not only in the Midwest but across the nation. In 1950, the Future Farmers of America Foundation sponsored a program comparable to 4-H's Farm and Home Safety Campaign. The FFA also rewarded exemplary achievement, providing cash prizes to clubs that made special contributions to improving local safety conditions. FFA foundation leaders presented two hundred separate fifty-dollar awards to winning chapters and medals to individuals making exceptional contributions. In 1950, an FFA chapter from Smith County, Tennessee, earned a particularly high recognition. John Farrar, an official for the Federal Security Agency, praised the chapter's members for implementing a variety of safety awareness activities, including constructing safety exhibits and distributing educational materials.[18]

The Smith County FFA chapter provides a vivid example of the type of vigorous effort a local youth club could mount. Members enthusiastically involved themselves in a variety of safety-related activities, including constructing exhibits, participating in Farm Safety Week, and enlisting the local media's involvement. Chapter members constructed an exhibit highlighting the most common farm hazards and showed it at their county and state fairs. Area youth also observed Farm Safety Week by visiting schools and private residences to raise safety awareness. They distributed safety materials to an estimated 85 percent of the county's citizens. Club members developed safety radio announcements and newspaper ads, which local media outlets featured. In addition, they educated themselves in first-aid techniques and distributed first-aid materials to area farmers. The Smith County example is representative of numerous other 4-H and FFA chapters throughout the nation and also demonstrates how successful agricultural educators were in galvanizing rural youth for the safety cause.[19]

Farm safety specialists, extension staff, and NSC officials assisted 4-H and FFA in mounting safety programming with a cooperative spirit and sense of civic responsibility, providing important professional advice, specific information, and help in designing educational programming. The NSC also played a crucial supporting role by providing educational materials, advising club leaders, and encouraging private companies to donate funds. FFA and 4-H members prided themselves on fostering a spirit of cooperation and collaboration as part of their organizational culture. For instance, the rising sun featured in the FFA symbol represented the hope of a "new day that will dawn when all farmers are educated and have learned to cooperate."[20] Such language echoes the rhetoric of many farm safety specialists who believed education and cooperation were crucial for reducing farm-related accidents. Such messages also encouraged farmers who embraced individualism to temper it with a cooperative attitude and openness to the latest approaches to farming. These ideals had much in common with the principles espoused by agricultural engineers and other reformers who had striven to modernize farming since the early twentieth century.

Private-sector donations were crucial to the success of 4-H and FFA safety programs. Benefactors included manufacturers of farm equipment, food processing companies, and even automobile manufacturers. Companies made such contributions not only to assist groups in their educational activities but to provide opportunities for improving public relations. Thomas Wessel and Marilyn Wessel, historians of the 4-H movement, indicated that Standard Oil's support of 4-H tractor maintenance programs largely stemmed from the company's desire to enhance its reputation among rural Americans, a goal confirmed by the long-standing nature of Standard Oil's support of these programs.[21] The nation's implement dealers also routinely donated tractors and provided the needed repairs for tractors used in youth tractor educational programs as a means of promoting positive relations with the communities they served.[22]

Corporate supporters of rural youth safety programming provided the financial assistance required for many accident prevention activities. The events included safety campaigns, recognition banquets, and the expenses required for trips to national youth meetings. The *Farm Safety Review* is filled with numerous examples of such corporate backing. In 1959, the Farm Equipment Institute provided the monetary support for the FFA's National Safe Corn Harvest Program. In 1960, the Oscar Mayer Company provided an award dinner for Wisconsin youth who had made special contributions to Wisconsin safety programs. In 1961, Allis-Chalmers and other agricultural equipment manufacturers paid for several 4-H Wisconsin youth members to attend the National Safety Congress meeting.[23]

Corporate donations sometimes also supported efforts to educate rural youth regarding the dangers associated with products used in farming. Throughout the 1960s, Firestone Tire and Rubber Company provided substantial monetary support for the 4-H Automotive Safety Program. Insurance companies likely contributed at least partially out of self-interest. For instance, in 1961 William P. Steinmetz, an assistant secretary to the National Association of Mutual Insurance Companies, indicated that his organization had faithfully sponsored the National Youth Fire Safety Contest. Steinmetz's association supported the program by developing contest materials, providing the monetary awards, and assisting in the contest's administration. In 1966, Schwinn Bicycle Company supported an effort to improve bicycle safety in Kansas and provided monetary achievement awards for the program. Instructors from Schwinn taught children about proper riding practices, basic traffic rules, and repair procedures.[24] General Motors Corporation was deeply involved in supporting 4-H safety efforts. In 1968, a *Farm Safety Review* writer stated that currently "516,600 4-H members were enrolled in the safety program, which has been sponsored by General Motors for the last twenty-three years."[25] GM contributed support to the cause in a variety of other ways. In 1961, it funded the development and distribution of a book entitled *4-H Leaders' Handbook,* outlining specific strategies and ideas useful in planning local 4-H accident prevention programs. GM's leadership also displayed personal interest in promoting safety. Company executives presented safety awards at the National 4-H Congress or national 4-H meetings, and in 1970, Richard C. Gerstenberg, GM's vice president, personally presented safety award winners with scholarships. Praising their accomplishments, he told recipients, "You are working in an area of vital importance to everybody. I am sure the 4-H Safety Program has saved many lives and prevented countless injuries."[26]

The corporate support that the nation's 4-H and FFA clubs and members received was well deserved. Youth could join both organizations, and in both they made substantial contributions to addressing safety problems. They initiated local safety awareness campaigns and participated in a wide range of other educational programming, including making posters, constructing exhibits, and producing plays. They also instigated more intensive efforts, such as personally examining local farms and correcting hazards. Youth sometimes even led community discussions regarding the issue, and they distributed educational materials throughout their communities, assisting rural Americans in gaining the latest safety information. Young people addressed the entire gamut of hazards, including those encountered at play, at work, and in the home.[27]

Agricultural fairs and meetings of various agricultural organizations provided a place to display the safety exhibits and posters the rural youths constructed. In 1948, James A. Schwalbach, a Wisconsin Extension employee, indicated that fifty-three Wisconsin youth had participated in "a farm safety poster contest" and that Maynard Coe had presented cash prizes to youth who produced exemplary work. The Allis-Chalmers Manufacturing Company granted the funds for the cash awards. Schwalbach emphasized the educational value of the poster competition, with event organizers distributing the winning poster throughout the state.[28] In 1950, a safety exhibit contest was held at the Pennsylvania Farm Show for the state's FFA youth, and the club that created the best exhibit received a monetary prize of fifty dollars.

The 4-H and FFA youth received guidance from club leaders, extension personnel, and *Farm Safety Review* writers on constructing effective visuals. In 1953, Samuel L. Horst, a safety education expert employed by the Pennsylvania Department of Public Instruction, while speaking at the NSC farm division meeting, provided a variety of suggestions concerning how such displays should be constructed: "Exhibits should be colorful and leave something in motion to attract attention. For motion you can use a continuous slide projector, poster machine, or mechanical devices such as a model of a power take-off, a corn picker or a tractor."[29] In 1957, Martha Kohl, a National Safety Council home economist, stressed the usefulness of such displays as a creative means to educate the public on a variety of safety issues. She stated that a good display should "fit the audience," have an "uncluttered" appearance, and possess appealing features to "attract attention" from onlookers. Kohl also advised that such artistic creations should possess a main idea and be well organized.[30]

Youth group leaders also incorporated drama in their programming, producing plays on accidents and prevention. These dramatic presentations provided an interactive approach to instructing youth and educating the larger community about safety. The plays' titles often made their message very clear; examples include *This May Happen to You, You Use Your Head: Else You Wind Up Dead,* and *Farm of Broken Men.*[31] The scripts of such plays were sometimes rather absurd. In *This May Happen to You,* the main characters were named Minor Injury, Permanent Injury, and Fatal. One of the Fatal's lines included the statement, "Yes, I am Fatal. I remove the victims who get beyond my brothers, Minor and Permanent." The script directs the actor playing Fatal to "laugh hideously" while delivering the line. Although safety plays might have had questionable artistic merit, they nonetheless served to educate both parents and their children. The popularity of such presentations diminished by the 1960s when they were no longer published in either the *Farm Safety Review* or the transactions of the National Safety Congress.[32]

Members of 4-H and FFA clubs often conducted "hazard hunts" in which they visited rural residences searching for hazards. Such investigations not only helped eliminate farm dangers, they also provided youth with useful educational experiences. Investigators visited local farms armed with safety checklists to assess the property for potential dangers. The examiner then alerted farm families to specific hazards and provided recommendations for correcting them. Participants usually returned to the farm to see if the problem had been corrected and sometimes even offered to fix the hazard. Such investigations also benefited the youthful investigators since they were required to learn about many dangers prior to conducting such investigations. Participants usually discovered a variety of hazards while conducting the hazard hunts, which sometimes involved investigations inside the home as well as outside on the farm. In 1950, Margarite McNally, a national 4-H leader, indicated that young peoples' investigations involved identifying the entire gamut of farmstead and household dangers, including those associated with electricity, livestock confinement practices, and agricultural equipment.[33] Such hazard hunts again reveal that some farmers permitted a high degree of intrusiveness from community youth. Outside experts would have been unlikely to receive such consent to inspect a farmer's property. The high value that many farmers placed on autonomy in their work was clearly displayed when OSHA sought to enforce greater safety regulations in the 1970s, for example.

Rural families were usually receptive to youth hazard hunts, however, since they were not conducted by outsiders, but by helpful neighbors. In 1962, Roger Moan, a 4-H club member from South Dakota, indicated that most of the people whose farmsteads he had examined were appreciative. Moan stated, "On the thirty-two farms that I inspected, I found people to be very cooperative and happy to learn that hazards existed." The large number of homes inspected in the program supports Moan's perception. The 4-H and FFA youth chapters were well suited to delivering the safety message due to their close connections to the local community.[34] Another factor to be remembered, however, is that they were delivering an educational not a regulatory message, leaving the farmers free to accept or ignore the advice.

A large number of hazard inspections related to the rural fire problem. For instance, in 1952, Iowa's Rural Fire Prevention Program included a particularly vigorous youth fire inspection campaign. Norval Wardle believed that the most significant part of this fire prevention program concerned youth's "inspection of their own and neighboring farms and the removal of as many hazards as possible." He celebrated Iowa's youth accomplishments, stating that "2,292 of the state's youth made 5,605 inspections, located 4,920 hazards and removed 2,802 of these dangers." Wardle also pointed to the 3,850 fire prevention discussions that youth had conducted with adults as another positive outcome of the program. During the same year, J. L.

Pennington, advisor of the Rugby, North Dakota, FFA chapter, also acknowledged the accomplishments of youth stating, "I know that the men of the Rugby Fire department agree when I say that the tremendous reduction in fire losses in our community is due almost entirely to the fine work of FFA boys."[35]

Fire was one of the earliest hazards rural youth organizations addressed due to the tremendous costs of fires in both financial terms and loss of human life. FFA and 4-H clubs began their fire education programs in the early 1940s. The NSC, the National Committee for Boys and Club Work, and the farm press often provided the required advisory assistance. However, the effort to reduce fires garnered the support of a diverse group of organizations. In 1958, the Volunteer Firemen's Association and various national insurance companies supported the development of New York's fire awareness program. Instructors for the project provided young participants with five informative sessions aimed at learning to identify and eliminate fire hazards. L. W. Knapp, Jr., Cornell University Extension agricultural engineer, stressed that fire awareness programs were crucial in educating youth about preventing and responding to fires and that this educational effort complemented fire inspection programs.[36]

The Iowa Farm Safety Committee directed another particularly successful statewide fire prevention program. In 1943, this effort took the form of the state's annual Rural Fire Prevention Program. The Iowa Association of Mutual Insurance Companies, Iowa Fire Prevention Association, Farmers Mutual Reinsurance Company, and radio station WHO provided the financial support for the annual event, and faculty members of Iowa State College assisted the state's school districts in administering the effort. The staffs of the state's rural school districts encouraged their pupils to engage in a variety of fire prevention activities, including encouraging local radio stations to disseminate the fire prevention message, participating in essay contests, and constructing prevention awareness exhibits. Norval Wardle stated that "their most important activities involved the inspection of their own and neighboring farms and the removal of as many hazards as possible." He also noted that "[d]uring the nine years this program has been underway, elementary school children from Iowa farms have removed 39,412 farm fire hazards. They have made a total of 61,705 inspections and uncovered 77,160 hazards. During the last four years, they have discussed fire hazards with 14,851 farmers and presented 366 fire prevention programs to 9,087 Iowa farm people."[37] Local youth clearly were able to win the trust of farm families in their own communities and thus to improve the safety of their surroundings.

Although fire was an important area of emphasis, farm technology was another area of great interest to rural youth organizations. Efforts included long-standing programs aimed at improving safe use of tractors and corn pickers, such as the 4-H's tractor maintenance program and the FFA Safe Corn Harvest Program. Youth organization leaders believed such training was crucial to provide young farmers with the tools to work safely in agriculture's increasingly mechanized environment. In 1949, Leon M. McNair, an official for the National Committee on Boys and Girls Club Work, indicated that such programs as the 4-H tractor maintenance program had been conceived to meet this need. Participants in the program were, he said, "instructed that a tractor which is improperly operated is a menace to life and limb, and that making a habit of the rules of safe tractor operation is equally important" as all other aspects of the care and operation of a tractor.[38] In addition, 4-H and FFA groups dedicated themselves to lowering the numbers of accidents involving other technologies not exclusively rural in nature, including automobiles and electrical appliances.

In 1944, the national leadership of 4-H established the tractor maintenance program to instruct the nation's youth in proper tractor use and care, a special concern as wartime labor shortages had resulted in less experienced drivers operating these dangerous machines. These views justified the program into the 1950s. In 1951, 4-H members from forty-six states participated in tractor maintenance programs, including clinics conducted by extension officials, 4-H leaders, or university faculty members. In 1951, a tractor training session held at Purdue University covered a variety of topics on routine tractor upkeep and proper safety guidelines.[39] Historians Thomas Wessel and Marilyn Wessel noted that every state but Pennsylvania had a tractor maintenance program under way by 1951.[40] The popularity of such instruction testifies to the reality that the tractor, a particularly important piece of agricultural equipment, was potentially one of the most dangerous.

Nebraska had a particularly robust tractor safety program with over eighty tractor clubs educating the state's youth in proper tractor operation. In 1954, William D. Lutes, a University of Nebraska farm safety specialist, stated that they are "doing a commendable job in developing a safety conscious attitude amongst young tractor drivers." Lutes stressed that clubs taught young tractor drivers about tractor "safety features" and the "hazards" inherent in improper operation and provided participants with opportunities to assess their operating skills in tractor-driving contests. In 1955, Henry Hatcher, the manager of the Omaha Safety Council, reported that Bennington, Nebraska, 4-H tractor-club members, "armed with sickles, scythes and corn knives," endured blazing one-hundred-degree summer heat to clear dangerous vegetation obstructions that could block drivers' vision at dangerous intersections. Hatcher noted that the entire community benefited from "unobstructed vision for the remainder of the summer."[41]

Tractor operator contests brought young drivers from across the nation together to highlight the need for responsible machine operation. In 1961, 4-H youth from forty states participated in two tractor operator contests: the Eleventh Annual Eastern United States Tractor Operators Contest and the Western United States 4-H Tractor Operators Contest. Tractor drivers from eastern states traveled to the Virginia State Fair held at Richmond, while their western counterparts gathered in Dallas, Texas. Participants were judged by a variety of criteria, including proper starting procedures, practicing sound driving habits, and employing correct refueling practices. The 4-H judges also took advantage of these contests to instruct parents in proper operating procedures. Such advice included avoiding the careless practices of having a second rider or wearing loose-fitting clothing that might become entangled in a power take-off.[42] These issues of safe tractor operation were major areas of concern long identified by farm safety specialists.

In the 1960s, the NSC and the Industrial Equipment Institute (IEI) worked together to standardize tractor safety education. In 1963, the NSC and the IEI cooperatively produced a tractor safety program for the nation's FFA chapters. A *Farm Safety Review* writer described the kit as "designed to assist any chapter in developing an effective local farm safety activity. It contains materials for a community tractor accident prevention program, among which are stickers and 'remember cards' given out during farm visits. Stickers and cards can be ordered free in quantity." By the early 1970s, these materials also included "a safety lesson outline, a large tractor safety poster and a guide for a community Slow-Moving Vehicle Program." Both the NSC and IEI received assistance from University of Nebraska agricultural engineers, who assisted in the development of the materials based on their research. In 1970, a *Farm Safety Review* article summarized the results: "During the 1969–1970 school year, more than 8,300 FFA chapters in forty-nine states use the tractor safety kit," the largest number ever to use such materials.[43] This program provides another example of the NSC's important role in developing and distributing safety materials to rural youth organizations that then enacted the programming.

Youth organizations also launched corn-picker safety campaigns to combat the frightening toll of injury and death caused by these dangerous harvesting machines. Crusades to reduce corn-picker accidents included conducting corn harvest accident surveys, using the media to raise awareness, and inundating farm families with safety information. The efforts to reduce corn-picker accidents were very similar to education efforts aimed at other rural safety issues in that they brought together a diverse group of rural organizations and institutions to enact a focused safety program. Their success was due to the cooperation among these organizations, which pooled their collective resources to combat a rural accident problem. Although corn-picking safety campaigns originated in the Corn Belt, other similar awareness crusades eventually reached a much wider geographical area.

In 1952, the Illinois FFA enacted the first corn-picker safety program, which informed thousands of farmers about the specific hazards of using these machines. The Illinois program received financial support from the state's implement dealers, the National Safety Council, the Illinois Farm Bureau, and other benefactors concerned about the problem. J. B. Adams, Illinois FFA president, planned the ambitious effort in which members of one hundred FFA chapters contacted ten thousand farmers with the corn-picker safety message. Art Johnson, an employee of the Illinois Farm Supply Company, celebrated the campaign's achievements, stating that youth "armed with leaflets, posters, film strips and pledge cards" had "launched a successful attack on corn-picker accidents."[44]

By 1954, the Illinois FFA corn-picker program had significantly expanded, with the campaign's organizers enlisting the aid of a total of "142 FFA chapters in the corn-picker safety program." The Newark, Illinois, FFA chapter's corn-picker campaign was typical of state efforts during the year. FFA leaders directed their members to each make five or six neighbors aware of the importance of corn-picker safety by distributing pamphlets, discussing some typical causes of accidents, and inviting local farmers to safety film showings. FFA members also placed safety posters throughout the community, asked local newspapers to carry corn-picker safety ads, and dispensed stickers for farmers to place on their machines displaying the promise to "pick corn safely."[45]

The Illinois program's success led to the development of a national effort. In 1957, the FFA leadership enacted the Future Farmers of America Safe Corn Harvest Program, representing a concerted midwestern effort to reduce picker accidents. The National Safety Council's farm division, the Farm Equipment Institute, and the National Retail Farm Equipment Association supported the expanded plan. In 1957, more than 2,100 FFA chapters from ten states participated. FFA members "made more than 150,000 personal calls on corn-picker owners" and dispensed "2,700 corn-picker safety information packets" and "161,250 stickers and pledge cards." FFA clubs whose contributions were exceptional were recognized at the FFA National Meeting.[46] The program was even popular in states beyond the Corn Belt. For instance, the contributions of the Berthoud, Colorado, FFA chapter illustrate the enthusiasm displayed by western rural youth. These young people "contacted 75 operators and visited 105 farmers to show them special safety precautions for the corn harvest season."[47]

In 1963, R. E. Hauptmann, a Vo-Ag teacher and FFA advisor from Mount Ayr, Iowa, discussed his club's experience with the Safe Corn Harvest Program at the NSC's farm division meeting. He revealed that Vo-Ag teachers throughout Iowa had consistently cooperated with each other to ensure the program's success and that the Mount Ayr community had vigorously supported the program, as evidenced by the local newspaper staff's willingness to publicize the event and the financial contributions of seven local businesses. Hauptmann also stressed that the FFA youth held each other accountable for carrying out the required duties: "Without my knowledge, class members were assuming the responsibility of talking with and encouraging the slow doers to get on the job and support the program. They let it be known that class members would not tolerate a poor job by any member." Such comments reveal that the young people's competitive nature and peer pressure worked in favor of strong participation in youth safety campaigns.[48]

Youth organizations in the Midwest and beyond also responded to dangers associated with electrical use, since electrification of the nation's farms posed special hazards. Rural youth responded by mounting safety awareness efforts devoted to the prevention of electrical accidents. In 1947, the Wisconsin Utilities Association assisted the state's 4-H club organization in conducting electrical safety seminars for the state's youth. Evelyn Evert, an official for the Wisconsin Electric Power Company, described the focus of the educational effort as various demonstrations highlighting the dangers of improper electrical wiring. Arkansas FFA youth were also active in promoting safe electrical practices through a 1948 program, supported by the Arkansas Power and Light Company, to prevent electrical accidents in their state. In the early 1950s, 4-H clubs in West Virginia also mounted electricity awareness programs, supported by home demonstration workers, county extension agents, and agricultural engineers.[49]

The ultimate effectiveness of these youth programs to promote safety awareness and reduce accident rates in their communities is difficult to ascertain. This is particularly true regarding the educational value of safety exhibits, posters, and plays. Although they probably did serve to instruct youth regarding a host of safety issues and to raise the public's safety awareness, it is difficult to tell to what degree. Even the era's 4-H and FFA club leaders admitted this point. In 1952, Richard M. Goss, a 4-H leader from Colorado, stated, "We can't tell exactly how many accidents we have prevented or lives we have saved. But we do know that many farms and home are safer places to live."[50] The 4-H historians Thomas Wessel and Marilyn Wessel also admitted the difficulty of assessing the effect safety education had in reducing hazards in the countryside.

Nevertheless, the NSC, state extension officials, and rural youth organizations collected enough favorable data to justify the continuance of rural youth safety programs. Wisconsin safety advocates, for example, believed young people had played a crucial role in reducing farm accidents; a brief 1952 article in the *Farm Safety Review* indicated that farm mishaps resulting in death had been reduced in the state from 168 in 1945 to 115 by 1951. The piece also revealed that every year in that six-year period had witnessed a gradual reduction in the fatality numbers. In 1955, Randall Swanson, Wisconsin's farm safety specialist, gave much of the credit to the state's youth stating, "We have had large reductions in farm accident deaths. Much of the credit for this goes to the youth organizations of our state."[51]

Although it is difficult to determine the overall effectiveness of hazard hunt activities, it is highly probable that they at least temporarily reduced the presence of dangerous conditions. Youth clubs that conducted such investigations often examined hundreds of farmsteads and identified hundreds of potential dangers. In 1951, the Rugby, North Dakota, FFA chapter inspected the properties of 496 farm families. Even though they did not correct the hazards themselves, they provided the families with specific advice on how to rectify the dangers. Investigations that included a second visit were probably more effective in reducing risks to farm families. An examination by a group of 4-H youth in Ohio featured such a structure and produced impressive results. In 1951, George Tewksberry, an employee of the Ohio Farm Bureau Insurance Company, stated, "After a second inspection tour, they found that Mr. Farmer had corrected over two thousand hazards, giving them a 53.5 percentage of hazards eliminated."[52] Club members who directly eliminated dangerous conditions ensured safety improvements. In 1954, the Bottineau, North Dakota, FFA youth club members revealed that they had "repaired or installed power take-off shields on 37 tractors, replaced 271 loose or broken tool handles, repaired or built 48 ladders, inspected 560 electrical appliances for shorts, repaired the cords on 189 appliances, removed the sharp edges from 540 machines, repaired floors on 50 hay racks and wagons." Youth groups who conducted hazard hunts throughout the era boasted of similarly impressive results.[53]

Rural safety leaders provided evidence to support the effectiveness of corn-picker safety campaigns, but they also noted the complications of assessing them. In 1953, Otto Steffey, a high-ranking official for the Illinois Farm Supply Company, credited his state's "Corn-Picker Safety Program" with reducing harvest accidents. Steffey noted that Illinois farmers had had 235 corn-picking accidents in 1951, a number reduced to 168 the following year. Robert Howey, a Vo-Ag teacher from Sycamore High School in Illinois granted that improved field conditions might have been part of the reason for the reduction in the number of accidents, but he added, "We like to think that part of it was due to our safety education program."[54] Minnesotans also experienced success with the campaign. In 1953, Glenn I. Prickett, a University of Minnesota farm safety specialist, remarked that Minnesota's Corn-Picker Safety Program had been worthwhile, stating, "One county reports results of this type of safety campaign: 'no corn-picker accidents in a heavy corn producing area for the last three years to October 17, 1953.'"[55]

The success of the corn-picker campaign is corroborated by both longitudinal and nationally grouped data. In 1955, Roland Espensheid, a Vo-Ag teacher from Franklin, Illinois, provided a more comprehensive picture of the initiative's positive effects in his state. He stated, "In 1951, we did not have an organized corn-picker program in Illinois, and we had, at that time, about 292 corn-picker accidents in our state. In 1952, we had 221; in 1953, we had 192; and in 1954, we had 115."[56] The FFA Safe Corn Harvest Program had expanded to ten states by 1958, so that thousands of farmers received information regarding the safe use of corn pickers.[57] Safety advocates believed that the campaign's dissemination had achieved significant cumulative results. In 1959, an anonymous *Farm Safety Review* writer indicated that just 83 corn-picker-related injuries occurred in 1958, reduced from 214 the previous year.[58]

Regardless of their ultimate impact, youth-oriented accident prevention programs represented some of the era's most ambitious safety education efforts. Safety advocates identified both 4-H and FFA as ideal organizational tools for reaching the rural public. Characteristics of youth safety programs such as their diverse nature, the large numbers of participants, and the favorable response of farm families justified this positive appraisal. In addition, such organizations' efforts to modify farm families' work habits fit well with the similar efforts of other organizations. For instance, the FFA Creed as published in the 1972 manual included the statement, "I believe in my own ability to work efficiently and think clearly, with such knowledge and skill as I can secure, and the ability of progressive agriculturalists to serve our own and the public interest in producing and marketing the product of our toil."[59] The words "efficiently" and "progressive" echo terms agricultural engineers and others promoting industrial agriculture had used for decades. Similarly, 4-H from its inception had been closely connected to the land-grant college system, which encouraged farm families to integrate a scientific and business ethos into farm operations.[60]

9

FROM COOPERATION TO DIVISION

BY 1960, THE NATION'S FARM SAFETY leaders could boast of having established a vibrant and coordinated national farm safety movement. In the following years, however, the effort's cooperative tone would fade. A variety of factors and events altered the movement's character, initially marked by war-inspired feelings of unity and patriotism. Farm safety leaders wished to increase their professional prestige and autonomy, which sometimes meant sacrificing harmonious relationships with other contributors to the movement. This change was made apparent by the creation in 1962 of the National Institute for Farm Safety (NIFS) as an organization separate from the National Safety Council (NSC). Another indication of the more contentious atmosphere was the increasing tendency for individuals involved in farm accidents to take legal recourse against farm equipment manufacturers for their injuries, resulting in growing numbers of product liability cases involving farm equipment.

By the 1970s, OSHA's attempt to improve agricultural safety was also sparking controversy as many farm community members resisted what they viewed as an unwarranted intrusion into their affairs.

AN AUTONOMOUS NATIONAL INSTITUTE FOR FARM SAFETY (NIFS)

Farm safety specialists separated themselves from the NSC and formed their own autonomous nonprofit organization, the National Institute for Farm Safety, in July 1962 at the yearly NIFS meeting held in Sarasota, Florida. The NSC's farm division had previously controlled these annual meetings, but farm safety experts wanted independence from the agenda of the larger organization. Farm safety specialists had long desired an independent organization. In a letter dated March 6, 1963, Norval Wardle, the first president of the NIFS, wrote to Howard Pyle, the head of the NSC, describing the origins of the new association. He wrote, "This is not a fly-by-night organization which developed overnight. It is the result of the desires, dreams, studies and plans of a large group of the professional workers in farm safety over the years since 1950 when I had the honor of first proposing it at one of our workshop get-togethers. It did not appear feasible at that time, but we think it is proper and right now."[1]

The correspondence between the leaders of the NSC and the newly independent NIFS also reveals that this move to enhance the professional prestige of farm safety experts caused some tension. In the spring of 1963, Wardle wrote another letter to Pyle revealing a lack of communication. He expressed dismay that NSC officials had not corresponded with NIFS officials and reiterated his determination to forge an independent organization. Wardle stated, "May I assure you again that we have every desire to cooperate with the National Safety Council but we do not plan to subordinate."[2] Although this particular letter was never sent, the correspondence between NIFS and NSC officials throughout 1963 alludes to the existence of some intra-organizational strain. On May 8, Norval Wardle sent a memo to Howard Pyle demonstrating NIFS members' desire to work harmoniously with NSC leadership. He stated, "May I assure you again that we have every desire to cooperate with the National Safety Council and particularly with the farm division. There is no desire or consideration of being in any way competitive but only complementary and supplementary."[3] The letter's conciliatory tone reveals the likelihood NSC officials had not greeted establishment of the NIFS favorably.

Additional correspondence between NSC and NIFS leadership provides more convincing evidence of a relationship in stress. In June 1963, Glenn I. Prickett, a University of Minnesota farm safety specialist, wrote to Randall C. Swanson, his University of Wisconsin counterpart, regarding the situation. Prickett indicated that he would be disappointed if any hard feelings with the NSC resulted from the NIFS's desire to be independent.[4] On November 15, 1963, Randall Swanson wrote to Norval Wardle: "All of us appreciate the fine work that you did as the first president of the National Institute for Farm Safety. It has been a difficult year because of resistance coming from the National Safety Council. I believe that we have conquered our major obstacles and that we can go forward from here." He wrote further that the organization's first meeting, held at Des Moines, Iowa, had been a "real 'touch and go' affair."[5]

Although the organization's first meeting resulted in some tension, Norval Wardle took the opportunity of his first presidential address to usher in a new period in the development of his occupation. He indicated that the creation of a professional organization meant "we are, as a group, now mature." Wardle alluded to the fact that farm safety efforts needed to be increasingly based upon sound research and analysis. He also indicated that a more rational approach would not only make the nation's farms safer but would also improve the stature of the farm safety profession, arguing: "Our safety programs can no longer be based on what we *think* is right. We have some knowledge, some facts in safety; here is where we should have our programs. As we get more and more facts, we can broaden and extend our programs."[6] Such words indicate that farm safety leaders now fully identified their cause as demanding a more systematic approach. They also appear to have fervently believed theirs was a challenge best solved by applying their unique brand of expertise to the farm safety problem. Such sentiments aligned with the view that many agricultural engineers had expressed for decades. Both agricultural engineers and farm safety advocates had consistently embraced the idea that farming, like other of the nation's industries, required rationalization and reorganization according to a new industrial ethos combining the latest technologies and new working methods.

NIFS leadership created more stringent membership qualifications, reflecting the desire to streamline the profession's identity and increase its recognition. Wardle emphasized that the organization would not only "promote the safety of farm people in all ways possible," but also "develop the profession of farm safety." In 1963, Randall Swanson concurred, stressing that the move to create a separate organization would serve to enhance his occupation's credibility.[7] NIFS policies limited full membership to individuals who devoted themselves exclusively to the safety profession. Such a move can most plausibly be explained as part of an overall effort to build a stronger professional identity and to provide those who devoted all their time to safety with controlling influence in the organization. Swanson expressed this opinion to C. E. Stevens, a safety specialist employed at the University of Missouri, stating that he did "not want to belong to an organization professing professional status on one hand and accepting anybody that had money for membership on the other." The organization's membership scheme allowed a candidate to become a full member only after being employed for at least five years as a safety specialist. Procedures evolved that required prospective members to submit detailed applications that were then reviewed by NIFS officials who would decide if the applicants' qualifications best suited them to be full, associate, or sustaining members.[8] This scrutiny indicated the increased emphasis on professionalism, the trademark of a fully developed occupational identity.

Associate members who worked less than full time on safety questions thus exerted less influence in the organization than did full members. In the summer of 1963, E. W. Foss, a Cornell University safety specialist, resented this practice, which in his view symbolized the dominance of a small group of individuals over the entire field. In a letter to Howard Pyle, he stated that upon attending a meeting of the incorporated group at the 1962 NSC Congress, "I learned that this newly formed institute had bylaws which restricted membership." He went on to state that this membership policy "automatically cuts out a large number of part-time farm or rural safety specialists."[9] The leaders of the newly formed NIFS did not view Foss's complaints with sympathy; rather, they seemed to fear that he represented a threat to the new organization. In an October letter from Randall Swanson to Norval Wardle, Swanson wrote, "You are aware, I am sure, of the Ed Foss situation which is rather disgusting, at best, but I am sure it will work out without any serious problem. We have a solid majority of the executive committee so that we can handle the situation even if some of them get foolish enough to support approval of such a move."[10] Extant letters do not provide sufficient evidence regarding what might have been Foss's retaliatory plan; such letters as do exist, however, indicate that veteran farm safety professionals wanted to limit the influence of those they perceived to be less qualified to lead the organization. Such moves can best be interpreted as an effort by the organization's leaders to solidify the strong professional identity that they hoped would provide them with greater influence and prestige.

Additional correspondence supports Foss's view that a select few were guiding the organization's aims and structure. In a letter from Randall Swanson to Norval Wardle regarding a meeting place for the 1964 institute, Swanson voiced concerns on the subject of holding their meeting at the University of California at Davis since it might threaten the institute's emerging structure. Swanson distrusted the intentions of that university's extension safety specialist stating, "Ralph Parks is not a member or a sympathizer with the Institute's organization." He also felt the NSC would have a prominent presence at the meeting and noted that "if this [meeting] were held in California, the Safety Council can travel to and from that area at their expense while the Institute will not have travel funds with which to protect themselves. I think it is very important that for at least another year the institute be held at a location where a strong full membership can protect the interests of our organization." Despite such controversies, most farm safety specialists confirmed their support for the organization: Almost all of them quickly joined the NIFS, revealing that the profession's nascent years had come to an end.[11]

DISSENT IN THE RANKS

The farm safety movement was also affected by a national trend of demanding greater manufacturer accountability and legal culpability in cases of accidental death or injury. In 1969, speakers at an Iowa Farm Safety Council meeting articulated such sentiments. Lex Hawkins, a Des Moines attorney who had represented auto accident victims, addressed the audience concerning the issue. Hawkins had also worked with Ralph Nader and other safety advocates in convincing the nation's political leaders to increase auto manufacturing safety standards. He warned farm equipment manufacturers of the dangers of ignoring the issue. He stated, "In my business as an attorney, I make my living off the manufacturers' and distributors' goofs in not protecting the user of machines and equipment." He went on to discuss how car manufacturers had already been forced to learn about the need to improve safety. Hawkins stated, "I saw what happened there . . . the industry refused to make cars safer so the government stepped in and drew up safety standards. This same thing can happen in farming." He also implied that the manufacturers had a lot to lose financially if they neglected the problem.[12] In 1974, L. R. Modlin, a John Deere attorney, expressed a similar view at a regional American Society of Agricultural Engineers (ASAE) meeting. He stated, "Fueled both by outspoken advocates such as Ralph Nader and by each consumer's own experience and frustration in coping with inadequacies and breakdowns of increasingly sophisticated goods available to the buying public, the consumer is demanding action and is getting it."[13] Such arguments indicate that the potential for legislation and the public's widespread demands for more safely engineered products were beginning to overshadow education as a means to improve farm safety. They also demonstrate the powerful influence gained by the consumer movement during the 1960s and 1970s. Historians such as Lizabeth Cohen comprehensively documented the powerful impact such campaigns had on American culture and politics. Cohen states that "through the 1960s and 1970s, the executive, legislative, and judicial branches of government oversaw the enactment of dozens of federal laws and regulations to protect consumers from harmful food, drugs, and cosmetics; unsafe manufactured products and vehicles of transportation; misleading labeling, packaging, and advertising," and a host of other areas in which consumers interacted with the private sector.[14] It is unsurprising that this changing political climate had begun to affect the agricultural sector.

In 1974, Frank Buckingham, an *Implement and Tractor* writer, also indicated that manufacturers and dealers were operating in a more challenging legal context. He stressed a number of changes in how judges interpreted the law that had expanded consumers' power. Such developments included reduced emphasis on operators being "free of fault" in accidents; the position that machines no longer needed to be the "sole cause" of mishaps but could be faulted if they provided merely a "contributory cause"; and the extension of written "warranties" to "implied" ones.[15] William Johnson, an NSC official, indicated that changing legal understandings made it more likely that accident victims would receive large monetary awards. He stated, "In some areas, the doctrine of punitive damages has been added. This means that if a manufacturer had acted wrongly you might almost say 'immorally,' as in not testing a drug, the courts will assess punitive damages. In such cases the penalties are usually large enough to impress upon the company the seriousness with which its actions are viewed."[16] In 1976, R. O. Diedrichs, an engineer who had served as an expert witness and had done extensive safety research, spoke at an ASAE winter meeting. He alerted his listeners to the expanding costs that liability case awards had exacted upon the nation's manufacturers. Such monetary penalties totaled $500 million in 1965 but had ballooned to approximately $50 billion by 1975.[17]

These expanding costs coincided with disagreements among members of the agricultural engineering community regarding how diligent farm equipment makers had been in incorporating safety in their machinery designs. The issue appears in some cases to have pitted the agricultural engineering community against itself. Those employed by private industry often believed that manufacturers were designing machines with safety in mind, while some working for public institutions such as universities often alleged that manufacturers had responded slowly. In 1980, the comments of Bill Field, ASAE Agricultural Safety Committee secretary, addressed this acrimoniousness: "There is a struggle within the farm safety profession between persons and groups being motivated by different concerns. Farmers, professors, insurance executives, engineers, and others are not all driven by the same interests and desires. Therefore, our joint efforts may be modified by the fear of adverse economic impact, hurting someone's feelings or the desire to avoid conflicts."[18] Field's statement appears to have described many of the elements affecting the farm safety movement as the dominant approach to resolving safety issues shifted from education to engineering and litigation.

Norval Wardle's accusations against agricultural equipment manufacturers were particularly visceral. In 1972, he spoke as part of a panel to an audience of farm safety experts at an NIFS summer meeting, along with safety specialists Carlton Zink and Randall Swanson, and commented on farm implement makers' accident prevention responsibilities. Although he believed some improvements had been made in agriculture-related technologies, he nevertheless thought that such efforts had been inadequate. Wardle stated, "The progress in actual machines, chemicals, electrical systems and farmstead equipment as used on the farm when compared to the knowledge in the field available to guide in this field is skimpy, kindergarten, inadequate and entirely unworthy [of the] mechanical genius of America." He also believed that the rise in liability cases was a positive development saying, "[o]ur courts and trial lawyers are performing a real function to make the manufacturers and merchandisers of such machines and equipment realize what the real cost of accidents resulting from such machines and equipment really is."[19] Wardle also expressed his belief on another occasion that farm equipment manufacturers treated accident prevention measures only as an afterthought.[20] This passionate safety advocate also came to believe that such views had cost him professionally.

In February 1978, Norval Wardle sent a rather emotional letter to Marvin Nicol, an NSC employee. Wardle wrote that he had suffered professionally from having aided individuals involved in lawsuits against agricultural equipment manufacturers. Wardle stated, "I have always felt that my first obligation as a safety worker was to the farm people and I have always tried to bring to light the truth of these accident cases. It has come to be that I am about the only one that they can turn to since Randall has stated that he will work only for manufacturers in such cases; and of course Edwin Tanquary and Carlton Zink, with their long association with manufacturers also work for them. As a result I have surely raised the ire of the manufacturers." He went on to explain what he viewed to be the sanctions of the professional community against him: "They have succeeded in getting me kicked out of ASAE and are still pushing me. What is one little guy doing opposing the great FIEI [Farm and Industrial Equipment Institute]? I never expected to do such a thing. The irony of life."[21] Although Wardle's situation appears not to have been exclusively related to his consulting work, his predicament does reveal that a contentious air had infiltrated the movement.

Wardle's correspondence with Wesley F. Buchele, a prominent Iowa State University agricultural engineer, further describes Wardle's problems. These documents reveal that the prominent farm safety specialist believed that farm manufacturers had been negligent in their duty to properly consider safety in their product designs. Buchele had requested information for a farm safety history he was thinking of writing. Wardle must have believed that such a history would place farm equipment manufacturers in a bad light, because he cautioned Buchele against the project, noting that his work with individuals suing farm equipment manufacturers had strained his relationship with farm equipment industry members. On May 25, 1978, Wardle wrote to Buchele: "Last August you wrote me a letter in regard to your project of writing a history of agricultural safety in the United States. Are you sure you want to do this?" Wardle appears to have thought hard about the idea, and he expressed his doubts about it to Buchele: "Who would publish such a story; not the ASAE, not the farm division of the National Safety Council. However, I have had offers from some people who think they know how to do it effectively. Personally, I have had no feelings about this but what about all of the farm people over America who are being sold out on their safety?"[22] Wardle described his recent troubles, including a suit against him brought by the Iowa Board of Engineering Examiners. He stated, "They were asking for a temporary and then a permanent injunction against me presenting myself in the state of Iowa as a registered professional engineer." Wardle denied the Iowa Board's allegations against him in his letter to Buchele and expressed his view that both International Harvester Company officials and other members of the FIEI had supported the Engineering Examiners' move. Wardle also believed the incident had hurt his relationship with the members of ISU's agricultural engineering department. He stated, "I have received the kiss of death as far as Ag. Engineers are concerned. I have dared to speak out in regard to the safety of farm machinery. I had rather thought the FIEI owned some, but not all."[23] Clearly, unlike many of his associates, Wardle believed that farm equipment manufacturers had been too slow to fully commit themselves to safety in their machine designs. Wardle also believed that his consulting work against agricultural equipment manufacturers had cost him professionally. Such correspondence reveals a movement characterized by conflict as much as or more than cooperation.

Nevertheless, despite these negative experiences, prior to his death Wardle appears to have at least partially reconciled with the proponents of the larger safety cause. In 1986, he was invited to attend the twenty-fifth anniversary NIFS meeting, which included a commemoration of the organization's establishment. Although he declined the request, his sentiments were positive. Wardle stated, "I am thrilled with the great progress which has been made in safety throughout rural America. Much progress has been made. We have safer homes, workplaces, equipment, highways and recreation places. The three great E's of Safety have all become more effective today, of Engineering, Enforcement, and Education. Let us put bickering and fault blaming behind us. Just work at all of these to see that each area is given proper attention and we will make still further progress through the years."[24]

Wardle was not alone in his view that manufacturers were not giving proper attention to accident prevention. In 1966, L. W. Knapp also believed that engineers had been too slow to include sufficient safety measures in their products. He thought that in some cases education had been overemphasized and had served as a justification for delaying other accident prevention measures. He noted, "It is possible that those of us involved in accident prevention efforts may be our own worst enemies. The common license of us all is to attempt, through education, to make individuals safer." He elaborated: "We have created and are perpetuating the idea in everyone's mind that accidents are an individual's own fault, thus neglecting the things that can be done mechanically and physically."[25] T. David McFarland, an NSC agricultural engineer, echoed Knapp's statements, stressing that "incorporating a high degree of safety into the product also lessens the need for extensive consumer education."[26] McFarland's comments reveal that the safety movement increasingly viewed education as an important but less dominant tool for reducing farm accidents.

Despite such disparaging comments, industry representatives often boasted of the many measures they had taken to integrate safety into their machine designs. In 1963, J. D. Morris, an Allis-Chalmers Manufacturing Company sales manager, pointed to a number of contributions his industry had made to the cause of safety. He pointed to improvements in tractors, such as electric starters to replace hand cranks, tail-lights on tractor-drawn implements, and a variety of features that made tractors more "comfortable" to drive.[27] A contributor to *Agricultural Engineers 1967 Yearbook*, an annual ASAE publication, listed "eighteen recommendations or standards" aimed at "operator safety." These included proposals for manufacturers to install safety features aimed at preventing falls, at protecting operators from moving parts, and at preventing misuse of operator controls.[28]

The individuals on the FIEI's safety committee also defended the industry's attention to safety. In 1969, the group's members issued a report maintaining that farm equipment manufacturers extensively tested their products for safety prior to releasing them into the marketplace. The writers also indicated that the organization was crucial in propagating safety standards, and they highlighted member companies' consistent support of farm safety education programs and the benefits of sponsoring safety awareness efforts. Engineers used such programs to gain worthwhile insights into potential equipment improvements, and operators received up-to-date safety information. The writers stressed that individuals who wrote equipment owners' manuals included a thorough consideration of safety in the booklets.[29] Such contrasting sentiments reveal that a divisive environment had significantly fragmented the movement, bringing agricultural engineers and farm safety experts into conflict.

In 1973, Wesley F. Buchele offered particularly insightful comments in a rough draft for a conference paper on the issue of manufacturers' response to safety concerns. Although he admitted that manufacturers could have been faster in implementing safety features in their equipment designs, Buchele was less visceral than Wardle in his view of the companies. Buchele indicated that, in his view, equipment makers did not respond slowly because they were evil or enjoyed the prospect of users being injured while using their machines. On the contrary, the delays could best be explained by the demands of the marketplace. Buchele stated, "Machinery manufacturers are shrewd businessmen; each has to balance the cost of manufacturing and marketing his product against competitors. Each does things that will make his machine most attractive in the marketplace. Low selling price is one of the more attractive items. Safety features may increase the cost and the complexity of the machine, increase maintenance problems, and reduce the potential numbers sold." Buchele went on to note that, despite the addition of safety features, operator injury was, in some cases, still possible while using farm equipment. Perhaps Buchele's view provides a reasonable middle ground between the era's two diverging camps of agricultural engineers.[30]

THE OCCUPATIONAL SAFETY AND HEALTH ACT

The enactment of the Occupational Safety and Health Act of 1970 and the subsequent creation of the Occupational Safety and Health Administration (OSHA) contributed to the increasingly contentious climate in the farm safety movement. The establishment of OSHA represented a significant expansion of the federal government's power to both institute and enforce safety regulations. Scholar Charles Noble characterized OSHA's creation as a significant shift in the history of U.S. regulation. Noble described the organization's goals and structure, writing "the new law centralized existing factory legislation by establishing a single federal agency with economy wide rulemaking powers. A federal inspectorate was established to monitor firm compliance and empowered to fine employers who violated standards."[31]

Despite its lofty aims, OSHA's involvement in improving safety in agriculture proved to be limited. In 1985, Glen H. Hetzel, who at that time was the NIFS president, expressed his disappointment regarding what he viewed to be OSHA's marginal impact during the previous decade: "Most of us have the occasional success that keeps us going and makes us try harder. Perhaps one of my biggest disappointments came when the OSHA regulations were, for all practical purposes, not applied to agriculture."[32] Hetzel's assessment of OSHA's minor impact appears to have been valid. During the early 1970s, most of the agency's regulations pertained only to farmers who hired employees; thus, farmers who depended on family members for labor continued to work largely outside its regulatory purview. A careful reading of the various standards in the *Federal Register* reveals this focus clearly: The only farmers required to obey standards were those who hired workers. Even in those cases, not many agricultural standards were imposed; the few instituted addressed tractor rollover protection, slow vehicle emblems, machinery guarding, field sanitation, and the handling of anhydrous ammonia.[33] Finally, the agency not only issued few regulatory standards for agriculture, it also devoted few resources to inspection. In 1976, Morton Corn, OSHA's assistant secretary, stressed that "very few of OSHA's inspections have been on agricultural sites," noting that "in 1975, OSHA conducted only 1,117 such inspections [in agricultural settings] out of a total of 100,000. This was 1.5 percent of OSHA's total federal inspection effort last year."[34]

Nevertheless, in a congressional hearing on OSHA's involvement in agriculture held in 1976, government officials and politicians expressed the widespread view that the agency had been overly intrusive. The hearing's participants leveled a range of criticisms regarding the agency's efforts to enforce its rather small set of agriculture standards. John Knebel, undersecretary of the U.S. Department of Agriculture (USDA), and Iowa congressman Charles Grassley were particularly passionate in their assessment of OSHA's overall mismanagement. Grassley complained that OSHA officials had failed to clearly communicate safety information and had worked too independently of other federal agencies. Most farmers, he went on, viewed OSHA officials primarily as "a roomful of lawyers drafting regulations which pertain to an area about which they know nothing." Knebel also believed the agency had done a poor job in explaining its standards, adding that it had not given farmers sufficient time to act on the regulations. Knebel pointed to the impracticality of regulating many rural work environments, which are both varied and decentralized, conditions that made education more suitable than regulation as a means to reduce accidents.[35]

These antagonisms may have been natural reactions to the additional responsibilities OSHA placed on individuals already struggling with a demanding occupation. In 1971, OSHA officials mailed instructional packets and paperwork to "farmers and ranchers" who employed non-family members on their farms. Employers were required to observe all the agency's safety standards and to "keep accurate records of work-related deaths, injuries, and illnesses, except those requiring minor first aid." Such record-keeping duties included making note of any significant exposures to potentially harmful substances and making yearly accident reports easily available to their workers.[36] It is easy to imagine that already overworked farm and ranch owners would resist these additional responsibilities as impositions on their limited time and energy. Such record-keeping requirements likely intensified the argument against regulating family farming operations.

In 1976, the editors of *IH Farm Forum*, a periodical sponsored by the International Harvester Company, surveyed their readers' opinions on OSHA. The survey was returned by only 829 *Farm Forum* subscribers, but respondents generally indicated they held an unfavorable view of the agency. Most believed that OSHA officials were largely incompetent and were wrongly interfering with farmers' work. Another common attitude expressed was that farm safety, like many aspects of farm work, was a family effort. Lloyd Ziegler, an Illinois farmer, reported his assumption that it was appropriate for adolescent boys to operate tractors. He stated, "I've trained three sons to operate tractors, and I started each one of them at twelve. First, they learned to operate the tractor with no equipment attached. Then, I let them operate the tractor with light loads for an hour or two at a time until they were ready to do bigger jobs. Any showing off and they were grounded." Ziegler was among the majority of the respondents who disapproved of what they viewed to be OSHA's misguided attempt to prevent accidents.[37]

Even before the International Harvester survey, farmers were clearly expressing their resistance to what they saw as OSHA's potential to threaten their work autonomy. In 1974, a proposal that would have required farmers to "retrofit" older equipment with safety features caused a visceral reaction from Harold V. Cobbs, a farmer from Lodi, Wisconsin, who was quoted as saying, "We will never submit to young, long-haired college kids from the Department of Labor inspecting our farms and slapping us with fines that could put us out of business."[38] In 1976, the rural public's reaction to a pamphlet entitled *Safety with Beef Cattle,* published by OSHA, was particularly harsh. Some felt the brochure lacked substance and was condescending in tone. Lyle Borg, a member of the Iowa Farm Bureau Federation, said, "Bureaucrats: we are not a group of helpless children. We are educated professionals operating the most involved, respected, and important industry in the world."[39] John Dvorak, a writer for the *Kansas City Times,* after investigating the issue, surmised that Missouri and Kansas farmers viewed OSHA's involvement in agriculture to be "less popular with farmers than a weed-infested-cornfield." Dvorak's interview with C. R. Johnson who farmed near Springfield, Missouri, and was the president of the Missouri Farm Bureau illustrated resistance to OSHA intervention. Johnson expressed the common sentiment: "There are just things that government can't do."[40] Such comments are in striking contrast to the enthusiastic response farmers made to safety programs proposed after World War II. The difference lay in the perceived attack on their independence, a valued aspect of farming as a career and a way of life. Earlier safety efforts, based on education, were seen as much less intrusive than the proposed new one, based on OSHA's imposition of federal regulations.

In March 1972, Dave Lucy, a contributor to *Farm and Power Equipment,* wrote a piece entitled "May I See You in Your Office?" expressing an adverse view of the agency. He described how intimidating it could be to have a U.S. Department of Labor inspector make an unannounced visit to a farm implement dealership. Lucy ominously told his readers that after a government employee tours your business, "Be ready for the boom to fall." He went on to describe an unpleasant experience that a Pennsylvania equipment dealer had had after an OSHA inspector's visit. The owner discovered that compliance with OSHA standards would cost $5,000. Changes required included updating the establishment's electrical system, "installing guard rails on all stairs," and renovating the shop's painting area to comply with ventilation standards.[41] Such stories, even those that didn't directly involve family farms, circulated throughout the countryside and undoubtedly contributed to increased fear and distrust regarding the consequences of federal attempts to improve safety.

Although large numbers of people in the agricultural community appear to have viewed OSHA's actions as misguided intrusions into their lives, such criticism was not entirely fair. For instance, the belief that the agency was grossly ill informed was probably overstated. OSHA officials arrived at their standards only after lengthy consultations with members of the agricultural community. For example, in 1972, during the organization's formative years, OSHA officials consulted an advisory panel convened to propose appropriate agricultural standards. This panel consisted of fifteen individuals representing a cross section of the nation's agricultural sector, including community group leaders, agricultural laborers, farm safety experts, and others with considerable knowledge of farming.[42] Taken together the authorities on these panels offered the benefit of years of theoretical and practical experience with agricultural technology. Inexperienced or overly zealous inspectors might have unfairly interpreted the resulting standards, but the standards nevertheless had been carefully considered prior to adoption. In 1973, panel member Rollin D. Schnieder stated, "One of the early criticisms of the standards was that 'they were poorly written and must have been written by those people in Washington.'" In fact, the standards were based on consensus standards adopted from associations or societies such as ASAE. In each instance, the American National Standards Institute or the National Fire Protection Association had adopted some of the industry standards and recommended them to OSHA. This was evident, for example, in the agricultural standards relating to use of slow-moving vehicle emblems and to handling anhydrous ammonia.[43] Schnieder's comments help balance the many negative reactions to OSHA's regulations.

The response to OSHA involvement in farming not only infused a more combative tone into the safety movement, it confirmed the persistent difficulty of attempting to improve farm safety through legislation. Farm safety leaders had for decades recognized the value farmers placed on their independent work environment. In 1968, a mere two years before enactment of the Occupational Safety and Health Act, Jack Burke, an NSC official, emphasized the differences between the farmer and the factory worker, stating that "no master mechanic routinely checks out a farmer's equipment for safety. There is nothing to keep him from running the most beat-up equipment that binder twine and baling wire can keep together. You can't get a ticket for plowing with a tacky old tractor." Since World War II, farm safety leaders had responded to this challenge by fashioning a collaborative and cooperative national education-based effort. By the 1970s, however, the involvement of an ever-increasing number of groups foregrounding their own distinctive interests created a tone far removed from the patriotic feelings and organizational cooperation that had given birth to the farm safety movement.[44]

ROLLOVER PROTECTIVE SYSTEMS AND FARM SAFETY

However unpopular the regulations, their paucity had an adverse impact on farm safety conditions. The story behind rollover protective systems (ROPS) illustrates the dramatic effects this situation had on the working conditions of the nation's farm families. ROPS is a particularly illuminating example, since prompt implementation of a relatively simple device could have potentially saved thousands of lives.[45] Large numbers of farm families used tractors without ROPS well into the 1990s, even though such devices had been in existence for decades. Safety organizations such as the NSC and others had advertised their benefits at least as early as the late 1960s.[46] Despite this, thousands of rural Americans continued to be harmed in rollover accidents throughout the late twentieth century. For instance, as late as 1988, 331 people on the nation's farms died in this type of accident.[47]

Observers had long recognized and publicized the significant danger of deaths from rollover accidents. As early as 1919, Edward R. Hewitt, a New York auto designer, identified the considerable ease with which a tractor driver could be harmed in a rollover. Over the next twenty years a number of other researchers also spoke up about the problem.[48] Not until the 1960s, however, was the problem nearly universally identified as a major safety issue. In 1963, a *Farm Safety Review* writer stated that "nearly two-thirds of the deaths" involving tractors were from rollover accidents.[49] In 1969, Jack Burke stated, "Of the one thousand annual tractor accident fatalities, about six hundred involve overturns. An estimated five thousand persons incur disabling injuries, often severe or crippling due to the nature of this type of accident." He went on to express his view that ROPS could prevent or minimize such catastrophic effects.[50] In the same year, University of Nebraska farm safety experts Rollin D. Schnieder and Robert J. Florell published a study supporting the use of such protective structures. Schnieder discovered that all Nebraskans killed in rollover accidents were driving tractors without ROPS.[51] A number of important health-related organizations also advocated ROPS as a safety measure. In 1967, the American Medical Association (AMA) issued a statement decreeing that all tractor makers should include ROPS: "The farm equipment industry [should] be encouraged to make available, as standard equipment, basic overturn protection of the operator on farm tractors that will conform to these standards."[52]

Throughout most of the era, both the technical expertise and effective designs for ROPS were available to the farm equipment industry. Inventors began submitting patents for tractor cabs and protective frames as early the second decade of the twentieth century. The safety functionality of these devices was disputed, however. Although some researchers deduced that safety considerations underlay the development and promotion of tractor cabs, critics of the early cab designs believed they were merely a means to improve comfort or protect the tractor. In 1988, Arnold B. Skromme presented a paper on the development of such devices to an audience of fellow agricultural engineers, expressing the view that safety had not been a goal: "It should be noted that none of these early inventors, or none of the several millions of farmers in the U.S. ever suggested that the safety structure be made high enough or strong enough to protect the operator."[53] Skromme's confidence that he understands the desires of all the nation's farmers regarding driver protection may be overstated, but it does appear his point is valid on some level. For instance, in 1923 Henry W. Hanson of Cheney Center, Colorado, received a patent for a cab with an application stating that the main purpose of his device was "providing a sun shade and a rain protector for the driver of the tractor."[54]

Despite the early controversy, by the 1950s both independent industrial designers and university researchers were actively developing devices to protect tractor drivers from rollovers.[55] By the 1960s, tractor manufacturers such as International Harvester and John Deere were also studying the effectiveness of various types of operator protection devices.[56] In 1967, David H. Bucher, a prominent member of John Deere's team of product designers, asserted that although not all new tractors were being fitted with ROPS, the company was testing the devices and would soon market tractors with them at a minimum price. Bucher emphasized his goal that tractor makers and users of rollover-protected tractors would work together to reduce deaths and injuries.[57] Nevertheless, Bucher stressed his view that use of this protective equipment should be "voluntary." Unfortunately, voluntary use of ROPS on family farms meant that tractor rollover fatalities persisted throughout the remainder of the century.

American companies were not alone in seeking solutions to the tractor overturn problem. Throughout this period the governments of many nations mandated that all new tractors be fitted with overturn protection device.[58] During the mid-1950s, for example, both Swedish and British engineers were testing various overturn protection concepts. In 1963, Carlton Zink commented on the Swedish government's efforts to regulate and standardize development of these safety devices: "In Sweden all new tractors must be equipped with protection for the driver. The National Swedish Testing Institute for Agricultural Machinery has developed a procedure for testing protective frames and cabs."[59] Some Northern European farmers thus generally benefited from ROPS protection long before their American counterparts. In 1968, E. C. Carlson, a high-ranking engineer for International Harvester, noted the contrast between European and U.S. practices regarding compliance with ROPS: "Compliance with the recommendation [in the United States] is strictly voluntary insofar as the manufacturer and the purchaser are concerned."[60]

Although few farmers and their families received protection from ROPS devices during the 1960s, this was not the case for many federal employees and some state workers. For instance, in 1960, state officials in North Dakota responded to the overturn problem by equipping all the state's tractors with ROPS, an action soon followed in other states.[61] For example, by 1965 most tractor drivers working for the state of Illinois had such protection.[62] Use of the protection also became common practice in the lumber industry. In 1963, Carlton Zink, a John Deere engineer and associate of the ASAE safety committee, wrote about the situation: "In some states where logging for lumber and pulpwood is important, there are legal requirements for a protective canopy over the top of the tractor and over the driver. Its purpose is twofold: to prevent falling branches or trees from striking the operator, and to protect him in case of a tractor overturn. The governing rules state that the canopy must be capable of supporting the tractor."[63] Thus, Corn Belt farmers remained an anomaly in their lack of mandated tractor protection throughout the twentieth and into the twenty-first century.

In 1995, J. R. Myers and K. A. Snyder, both prominent safety experts, revealed that as late as the early 1990s, large numbers of farmers were still operating tractors that had not been fitted with ROPS: "In 1993, over 4.8 million tractors were in use on U.S farms, of which 2.98 million lacked ROPS." The authors indicated that older tractors still in use and lacking the device could be fitted with one for $947, a small amount considering the high emotional and financial cost of the loss of a human life.[64] For international comparison, in 1967, Sverker P. E. Persson, writing in *Agricultural Engineering,* reported, "More than half of all tractors in Sweden now have safety frames or cabs."[65] This high figure was achieved because new tractors were required to have ROPS and the nation's leaders had instigated a vigorous retrofitting campaign. By 1970, farmers in the United Kingdom also benefited from such protections, since ROPS were mandatory on tractors sold there.[66]

Many American tractor drivers thus remained without ROPS because manufacturers were not required to supply the device and large-scale efforts to encourage retrofitting older tractors were never instigated. The major tractor makers voluntarily agreed in 1985 to make ROPS standard on all their tractors, but many farmers continued to use older tractors without protection.[67] A contributing cause of this lack of progress lay in the attitude of many agricultural engineers, who believed that the decision to use ROPS should be left to individual farmers. In 1966, Merlin Hansen, a prominent ASAE member, stated, "It is hoped that the combined efforts of everyone on a voluntary basis will substantially reduce tractor accident and fatality rates. By voluntarily providing effective protective devices and by intensively publicizing their use, the farm machinery industry can effectively perform a vital humanitarian service."[68] In October 1968, International Harvester engineer Ernest C. Carlson acknowledged that ROPS use significantly reduced the chances of death or injury from tractor rollovers, but he maintained it should be the farmer's choice whether to use rollover protection.[69] This emphasis on voluntary compliance undoubtedly contributed to the nation's large numbers of deaths and injuries from tractor upsets.

ALTHOUGH FARM SAFETY LEADERS had initially emphasized cooperation and coordination of efforts, a different tone eventually dominated. Controversies surrounded the creation of the NIFS, OSHA's efforts to improve agricultural safety, and the turn to lawsuits to address farm equipment accidents. Farm safety specialists who wished to establish their own organization, control who entered their profession, and market their expertise sometimes displayed more self-interest than cooperation. However, efforts to enhance professional authority are understandable in light of the growing emphasis on expertise throughout American society. OSHA officials' efforts to improve farm safety met with resistance and suspicion from the farming population and the safety community, with most doubting that a regulatory approach could resolve the agricultural safety problem. Finally, the expanding number of lawsuits led to severe financial penalties for farm equipment manufacturers that neglected safety features. Even though members of the agribusiness community, farm safety professionals, and farm families themselves cooperated in funding, initiating, and participating in farm safety efforts, fragmentation and divisiveness had infected the movement by the late 1960s. The cooperative tone that had characterized the movement during the Second World War had diminished, if not vanished, by the 1970s.

10

CONCLUSION

FARMWORK IS HAZARDOUS WORK, and it always has been, for a variety of reasons. The living and working conditions of Corn Belt farm families during the mid-twentieth century were undoubtedly complicated and unique in American society of the time. Farmsteads functioned as both home and workplace, whereas most Americans were by then living in urban areas and worked in places clearly separated from their homes. This distinction was very present both to individuals who had spent their entire lives in agriculture and to those raised on farms but who had left rural America for the cities. Jim Klein summarized the differences between farmwork and other types of work this way: "There are so many more variables. You are not on a factory floor where variables such as light and weather could be better controlled."[1]

Other factors contributed to farm families' vulnerability to accidents. Farmwork is seasonal, presenting the special health risks associated with working in the open under any and all weather conditions. Farmer John Van Ringelestein emphasized the cyclic stresses of farming, noting that "you'd work like dickens on days you had to get the crop in, and there were times—rainy days—you could take off."[2] During the research for this book, many farmers offered similar assessments of the intermittently frenzied pace and seasonal stresses of their work, and they spoke as well of its intense economic pressures. Financial demands and uncertainties contributed to farmers' enthusiasm for adopting the latest machines, chemicals, and buildings in their efforts to improve their competitiveness. Unfortunately, many of these agents of industrialized agriculture hid hazards that could contribute to devastating injuries or even death.

The farm safety movement emerged to help farmers address these issues. It admittedly had some successes, although the degree of its effectiveness in reducing farm accidents is difficult to determine. Safety activists often measured the success of their education efforts by such superficial means as counting the number of safety committees established, the numbers of attendees at events, or the quantity of safety materials distributed. Safety advocates engaged in few systematic appraisals of their activities, which helps to explain this ambiguity. Much of the numerical data supporting educational programs was anecdotal in nature rather than the result of carefully conceived studies. This situation should not come as a surprise, since farm safety activists appear to have been sufficiently challenged by their attempts to gain statistical understanding of the numbers, types, and causes of the accidents themselves. Nevertheless, safety advocates did occasionally provide state-level data demonstrating reductions in some kinds of accidents following the enactment of specific safety educational programs.

Although ascertaining the overall effectiveness of safety education programs presented difficulties, some conclusions can be drawn. Farm safety leaders successfully achieved high levels of participation, developed a diverse range of activities, and expanded their programming throughout the era. Thousands of rural Americans participated in safety activities. Youth organizations, such as 4-H and FFA, were committed to safety programs throughout the era, reaching thousands with their message. Safety leaders also proved particularly effective at both identifying farm hazards and responding to emerging issues by quickly fashioning safety programs uniquely suited to specific areas of risk. They were able to do this successfully in the context of a range of accident risks, including farm equipment, agricultural chemicals, and a myriad of home and farm electrical devices. Their efforts were also directed at averting mishaps with livestock, avoiding farm fires, and preventing recreational hazards. They learned to utilize a variety of media, including print media, radio, and television, in their efforts to improve farm families' health and welfare. In addition, rural safety activists demonstrated a high degree of pedagogical experimentation throughout the era, utilizing full-scale demonstrations, models, and competitive events to inspire people to live and work more carefully. Farm safety leaders clearly had great success in mounting a vibrant safety movement that provided effective safety education for the nation's farm population.

Despite the farm safety movement's accomplishments, however, problems persisted. In the 1980s, significant numbers of agricultural engineers, health professionals, and journalists continually decried the large numbers of people killed or injured on farms. In 1988, Kelly J. Donham, a faculty member at the University of Iowa's Institute of Agricultural Medicine and Occupational Medicine, stressed that much was still to be done to meet the tremendous challenge of improving agricultural safety and that government farm agencies, ag-related businesses, and various other farm-related groups needed to devote more resources to this important rural issue. Jim Leach, Republican congressman from Iowa, echoed Donham's recognition of the seriousness of the problem stating, "Romantically, we think of agriculture as the finest way of life in the world—unfortunately we know it's not the safest."[3]

The persistence of the farm safety issues highlights the seemingly intractable nature of preventing farm accidents. Historian Thomas P. Hughes in his book *American Genesis: A Century of Invention and Technological Enthusiasm, 1870–1970* investigates and analyzes a century of technological innovation and exuberance that he terms "the American Genesis."[4] Hughes states that in his book he "presents practitioners of technology confronting insolvable issues, making mistakes, and causing controversies and failures. It shows the practitioners creating new problems as they solve old ones. This book intends to present the history of modern technology and society in all its vital, messy complexity."[5] These comments could apply as well to the farm safety story. Agricultural engineers, chemists, and other experts successfully produced machines and substances that aided farmers in dramatically increasing production, but the innovators were unable to control or mitigate all the unintended consequences of these technologies. Costs included a variety of accident risks and health hazards, threats that persisted long after the introduction of gasoline-powered tractors, farm chemicals, and other instruments of industrialized agriculture.

Critics of farm safety efforts provided convincing evidence of the need to place greater priority on families' well-being. For instance, in 1988, the *Daily Dispatch*, a Moline, Illinois, newspaper, featured an article by journalist Angella Herrin on her investigation into the farm safety issue. Herrin wrote, "In 1983, the federal government spent about $4.34 per U.S. worker for on job safety programs but only thirty cents per farmer." Such financial disparities were noted by other individuals throughout the era. Kelly J. Donham agreed that funding was lacking and pointed out that many nations were ahead of the United

States in addressing the farm safety issue.[6] In 1989, Sidney M. Wolfe, editor of the *Health Letter*, a publication of the Public Citizen Health Research Group, supported these perspectives: "The reality that most farms are family units makes the safety regulations which do exist irrelevant. Congress had forbidden OSHA funds to inspect farms of less than eleven workers: this essentially means inspections for dangerous machinery and practices are forbidden on U.S. farms. Sadly, OSHA is not using its authority to inspect large farms: in 1987, it inspected a total of thirty-two large farms—only after a death in each case!"[7]

Some agricultural engineers and safety specialists believed that the farm equipment industry could have done more to safeguard farm families by moving more quickly to introduce safety features into their machinery designs. The relatively slow progress regarding adoption of rollover protective systems (ROPS) in the United States provides a particularly illuminating example. Seasoned safety experts such as Norval Wardle repeatedly expressed their view that the nation's agricultural manufacturers had moved too slowly in improving the safety of their equipment designs. Arnold B. Skromme, speaking in 1990, revealed the persistence of the problem. Presenting statistics at the ASAE annual meeting showing that American farm families experienced considerably more deaths and injuries than did their European counterparts, Skromme went on to summarize his conclusions: "An historian uncovers a strange situation in the United States—although this country had led the rest of the world in new inventions, such as the first pickup hay balers and forage choppers, the first all-purpose farm tractor, the first vertical silos, the invention of hybrid seed, etc.—this country has failed to take similar leading action to design machines and buildings to prevent accidents." The entire agricultural community, not just the farm equipment industry, had often placed greater priority on economic than on safety considerations.[8] Emphasis on farm machinery design improvements as a means to enhance safety accords with a late twentieth-century trend across many industries and areas of American life described by historian John C. Burnham: "[P]hysical or engineering means overwhelmingly came to dominate safety, obscuring even the continuing programs to educate people to behave so as to avoid accidents."[9]

Farmers interviewed for this project were generally less critical of farm equipment manufacturers than were the experts. Although many acknowledged that the era's agricultural technologies could have been designed with greater attention to safety, they also believed such improvements took time. Farmer Don Rimathe agreed that mid-twentieth-century machinery was less safe than later models, but he emphasized fatigue as another important contributing factor in accidents.[10] Pat Larkin, also a farmer, saw safety as a matter of personal responsibility: "I suppose they could have done more." He went on to say that he never viewed farming as more hazardous than any other job and that the dangers were "just part of it."[11]

Farmers recognized the hazards of farming, but they did not dominate their thinking about their work. Thus, farmers participating in this study did not downplay the very real hazards of their profession, but they expressed a range of opinions based on their individual experiences. Rex Peterson recalled the first time he let his son drive the tractor in the field without his supervision. He stated, "I could hardly stand it; I was so worried. It just bothered me that I couldn't see him."[12] Peterson grimly recognized the importance to accident prevention of constant vigilance since his dad had lost an arm in a corn-picker accident decades before. Another farmer, James Wilson, reflected with a matter-of-fact attitude on his many years in farming that "those were the risks you took. I was just lucky. It was more luck than sense; accidents happen and they happen quickly."[13]

Such comments reflect the general feeling among farmers that although farming presented hazards, personal caution was the best approach to avoiding injury. Many farmers of the era also expressed either skepticism or outright hostility to greater regulation of their farms, opinions strongly voiced in relation to OSHA regulations promulgated in the 1970s. Farmers may have been enthusiastic about adopting new technologies and participating in some forms of agricultural education, but they still viewed themselves as independent entrepreneurs who foremost trusted their own instincts about farmwork.

Interviewees also articulated the view that farmers were often more concerned about cost than about safety features. Lawrence Schmitz expressed the opinion that farmers of an earlier era desired "power" and were less than enthusiastic about "fancy" attachments.[14] Larry Glenn remembered appreciating his first tractor cab more for keeping him comfortable and out of inclement weather than for its safety benefits.[15] Wesley Longnecker also believed that price was a crucial element in the acceptance of safety features, stating that additional features mean "it's going to cost more."[16] In light of the economic pressures farmers faced in terms of the rising costs of equipment, seed, and chemicals, it is unsurprising that they invariably made price-conscious decisions about major equipment purchases.

THE FARM ACCIDENT STORY provides insights into the era's transformations in rural life. Although sophisticated technology assisted farmers in increasing productivity, reducing the drudgery of agricultural work, and bringing modern conveniences into farm homes, such modifications also fundamentally altered the farm family's living and working environment. The era witnessed the introduction of a host of new types of equipment and chemicals that if used improperly could cause catastrophic injury or death. By the mid-twentieth century, rural Americans had recognized the transformation and its relevance for the farm family. In 1955, Melvin Simmons, member of the Jasper, Missouri, FFA chapter, noted the gulf between the farmers' potentially dangerous environment and the public's idyllic image of farm life, stating that "city people consider the farm a place of safety and quietness." By the early 1970s, high-level government officials had also acknowledged the risks of technological innovation. During his presidential proclamation of Farm Safety Week in 1973, President Richard Nixon stated, "The unfailing supply of food and fiber provided by the nation's largest industry, agriculture, has been the foundation of American prosperity since our country's beginnings. Abundance on the farm has, in turn, been stimulated by constant technological progress. But the blessings of technology sometimes have been mixed, as each advance has also brought a new potential for injury."[17]

The farm safety story also reveals farmers' continued adherence to the principle of independence and acceptance of their occupation's entrepreneurial nature. This autonomy, however, could make farmers more vulnerable to accidents than those who labored in other occupations subject to greater oversight and regulation. Farmers had to rely on outside individuals to improve safety conditions. Farm safety experts devoted their entire professional lives to investigating the nature of the farm accident problem, proposing technical solutions, and fashioning educational programming to help reduce risks. Thus, farmers not only depended on their own judgment regarding proper work practices, they also counted on the work and advice of these professionals. Agricultural engineers, farm safety experts, and extension personnel labored to identify the specific dangers of an increasingly complex and changing rural environment. They funneled their solutions through an organizational network of governmental and private institutions. The farmers' dependence on the experts for safety solutions echoes their reliance on other outside groups: financial institutions for credit, government agencies for farm subsidies, and fertilizer companies to maintain the soil fertility. Although agricultural policy makers had promoted reliance on farm safety experts to cope with agricultural change since the early decades of the twentieth century, the creation and launch of a farm safety movement required the stimulus of a wartime labor emergency. Taken together, these developments show that by the mid-twentieth century farmers were probably less independent than they imagined.

Finally, this work reconfirms the increasing influence through the mid-twentieth century of an entire range of outside experts who marketed their knowledge to meet a variety of farm safety challenges. Even though ever fewer families could make a living on the land, professions associated with agriculture were actually growing. In 1970, Dr. James Davis, the chancellor of the University of California at Davis, commented on the many opportunities in agriculture, stating that "[m]any today are in agri-business—industries such as machinery or chemicals with farms as a market. Others are in rural sociology or environmental toxicology or one of the many other fields that colleges have had to add to their curriculums to keep pace with the agricultural revolution."[18] The emergence of the occupation of farm safety specialist, devoted to modifying farmers' work habits, supports such comments. This expansion mirrored the growing influence of specialists in many areas of American life and the tensions that such changes created for those seeking to maintain their autonomy.

EPILOGUE

BY 2016, MANY CORN BELT farm families planted their fields with genetically modified seeds, drove tractors fitted with Global Positioning Systems, and used a diverse mix of computer applications. Many families expand their operations as farming continues to grow in scale.[1] Technologies such as self-driving tractors and drones will mostly likely make farming even more unrecognizable from what it was a hundred years ago.[2] In 2010, Mark Stutsman, a precision agronomist, stated, "It won't be long before an Iowa corn farmer will plant and harvest between 25,000 and 40,000 acres annually, up from 350 acres today."[3] Although his comments may prove to be overstated, they confirm the ever-present reality of unrelenting change.

Farm families still encounter many hazards, but much progress has been made in recent decades to reduce their risks.[4] A National Institute for Occupational Safety and Health (NIOSH) study conducted between 1982 and 1996 found that youth farm fatalities had decreased. The study concluded that a smaller farm population, engineering improvements, and enhanced emergency technology contributed to the drop.[5] In 2008, Charles Schwab, an Iowa State professor of agricultural and biosystems engineering, stressed the hazard reduction benefits of features such as automatic shutoff devices and tractors with rollover protection.[6] In 2015, Aaron Dufelmeier, an Illinois State Extension director, concurred, noting that manufacturers had made significant improvements.[7] Clearly decades of efforts to reduce hazards have not been in vain.

The cause of farm safety continues to attract a diverse group of organizations and professionals. The National Safety Council, farm youth organizations (4-H and Future Farmers of America), and state extension staff are all deeply involved. In addition, a host of other organizations have more recently moved to address the cause; the Centers for Agricultural Disease and Injury Research Education, AgrAbility, the National Farm Medicine Center, the Carle Foundation Hospital Center for Rural Health, and the Childhood Agricultural Safety

Network all work to bolster safety efforts.[8] Kevin Barlow, a journalist for the Bloomington, Illinois, *Pantagraph,* reported that a recent meeting of the International Society for Agricultural Safety and Health had attracted "engineers, educators, insurers, physicians, nurses, veterinarians, statisticians, communicators, anthropologists, and business leaders."[9] This impressive mix of specialties confirms the primacy of a multidisciplinary approach to safety.

The corporate sector continues to donate monetarily to safety efforts, with Farm Safety For Just Kids listing the 2016 supporters of its education efforts as including AGCO corporation, Cargill, Deere and Company, among many others.[10] On the other hand, critics still argue that such resources are often inadequate and that government support, too, is often insufficient. This affects funding and staffing of important resources; for example, the University of Minnesota Extension Service did not fund a farm safety coordinator from 2008 to 2015, and some Minnesotans blamed lack of resources for an uptick in farm accidents between 2003 and 2013.[11] These grievances echo the concerns of farm safety advocates in the mid-twentieth century.

Significant improvements have been made in researching the safety issue, however, largely due to the establishment of ten Centers for Agricultural Disease and Injury Research Education and Prevention. These organizations, created by congressional order in 1990 as a cooperative initiative involving the Centers for Disease Control and Prevention (CDC) and NIOSH,[12] operate within NIOSH and with NIOSH support to create new solutions for agricultural health problems.[13]

Some of these institutions, such as the National Farm Medicine Center of Marshfield, Wisconsin, predate establishment of the program. In 2005, Dean Emanuel, a cardiologist and the Center's first director, explained that the Center's original mission had been to treat "farmers' lung disease." Nonetheless, Center physicians soon began to research a variety of ailments afflicting farmers, including helping accident victims who had lost limbs, treating osteoarthritis, and even investigating hearing loss. By the 2000s, the Center had treated thousands of farmers, trained emergency responders in treating farm accident victims, and opened up new areas of research. For instance, in the early 1990s, it began addressing the rural cancer problem, establishing the Rural Cancer Research Center.[14]

AgrAbility, a federally sponsored program established in the 1990s, also treats a more diverse set of health issues affecting rural residents.[15]

The graying of the farm population makes AgrAbility's focus on rehabilitation particularly crucial. In 1982, the average age of an American farmer was fifty, but the figure had risen to fifty-eight thirty years later. Rural health advocates are especially concerned about older farmers, who are generally more vulnerable to serious accidents and suffer from accumulated physical wear and tear brought on by their profession.[16] AgrAbility has contributed greatly to alleviating farmers' suffering from the aftereffects of accidents or the accumulated damage of years of manual work. In Wisconsin, for example, AgrAbility provided services and equipment to approximately six thousand farmers between 1991 and 2015.[17]

David Travis, a sixty-seven-year-old dairy farmer, was among those who benefited from AgrAbility's rehabilitation and environmental modification services. Travis suffered from several ailments, including arthritis and back problems, but he continued to work: "I guess once this is in your blood, it's in your blood." David W. Kammel, a University of Wisconsin agricultural engineering professor, referred Travis to AgrAbility. The organization's staff evaluated Travis's situation and then provided him with mats and heated floors for his milk house and made other adjustments to ease his work.[18] The story of Scott Collier, a farmer in the neighboring state of Minnesota, also showcases AgrAbility's innovative approach. Collier had suffered a spine injury in a serious auger accident. AgrAbility provided Collier with assistive equipment valued at $68,000, including a device to lift him into his combine and a wheelchair fitted with "track wheels." This equipment allowed him to continue harvesting his crops. These farmers' experiences illustrate the great potential of the rehabilitative approach for addressing agricultural health issues.[19]

Despite these encouraging developments, the causes of farm injuries and fatalities remain similar to those in the past. Farm families still encounter hazards associated with machines, chemicals, and livestock. Farmers continue to be warned about the potential for electrical accidents, just as their predecessors were, and they still live and work far distant from emergency resources, just as in the past.[20] Farm safety experts also acknowledge that farmers' work culture still poses challenges. In 2005, Dr. Dean Emanuel stated, "This is a highly independent population we're dealing with, and they just don't run to the doctor every time they have something wrong with them."[21] Safety advocates have made the point for decades: Some farmers occasionally place a higher priority on their work than on their health.

Another unfortunate historical continuity relates to the persistence of tractor rollover accidents. NIOSH investigators found that of the 516 farm fatalities occurring nationally between 1992 and 2005, 101 involved tractor rollovers. In 2007, Iowa State professor Charles Schwab supported the findings, stating that "tractors are the number one killer of farmers in Iowa and in the nation and have been for years."[22] The United States continues to lag behind many European countries in promulgating regulations requiring rollover protection on farm tractors. Some American farmers continue to operate tractors without ROPS, sometimes resulting in accidents causing injury and death.[23]

Many farm safety supporters find the tractor rollover problem to be particularly exasperating. As late as 1992, four million tractors nationwide continued in use without ROPS.[24] Nonetheless, in some states, such as Iowa, deaths in rollover accidents declined in the early 2000s, providing reason for optimism as the trend strengthens of replacing old unprotected tractors with newer ones with roll-bar protection.[25] In 2015, Matt Keifer, the head administrator of the National Farm Medicine Center, voiced his frustration that any tractors without ROPS were still in use, stating, "I don't know why it's taking so long for us to figure this out."[26] The issue has been known—and a solution for it has been available—for over half a century.[27]

Farm youth also continue to be outside the purview of most governmental regulations, and some believe that this situation has contributed to higher than necessary accident rates on family farms.[28] In 2007, Barbara Marlenga, a farm accident researcher and safety advocate, stated that "removing the family exemption from Hazardous Occupations Orders and raising the age restriction for performing hazardous agricultural work from sixteen to eighteen years would be efficacious in preventing the most serious injuries experienced by family farmworkers."[29] Unfortunately for proponents of a legislative approach, efforts to provide additional regulatory protections for farm youth continue to fail, just as had earlier efforts in the 1970s.

In response to the call for greater governmental involvement, the Department of Labor (DOL) proposed a law preventing youths age sixteen and under from doing some farm tasks, such as climbing high structures, inoculating livestock, and operating ATVs. Although the law would not affect children working on farms owned by their immediate family, it would affect their ability to work on property owned by members of their extended family.[30] The proposal garnered widespread criticism especially because it limited young people's activities on farms owned by uncles, aunts, or grandparents. Kristi Ruth, a young woman who not only grew up on her parents' farm in Columbia, Iowa, but had also suffered a serious injury there, spoke against the measure: "It's shocking to me that the government thinks it can fix a problem with a piece of paper. You have to encourage people to want to do it themselves." Iowa governor Terry Branstad also opposed the measure, calling it "a prime example of federal overreach."[31]

In the spring of 2012, the DOL capitulated to public hostility, and the measure was defeated.[32] Farmers' reasons for opposing it were similar to those given when opposing OSHA regulations in the 1970s. They argued, for example, that extended family, such as nieces and nephews, often represented an important labor source and that such work experiences provided youth with valuable skills. John Juusola, a farmer from Darwin, Minnesota, expressed his opposition, stating that the defeat of the proposal was "good news for kids" because "they can get out and get some experience."[33] Reid Maki, an official for the Child Labor Coalition and a supporter of the measure, characterized the debate: "There was tremendous heat, and I don't think it helped that it was an election year."[34] The response to these proposed child-protection regulations confirms that the farming work culture differed considerably from that in other economic sectors, such mining and manufacturing.

Yet this work culture can be a source of strength. Even as Corn Belt farmers cope with a world that buffets them with the upheaval of economic stress and technological change, their sense of communal responsibility remains constant. In fall of 2014, the *Daily Gazette*, a newspaper in Sterling, Illinois, featured a story about a young, local farm family beset with tragic circumstances. Brad Shippert had contracted meningitis while his wife was fighting cancer. Seven farmers responded, relieving him of the burden of harvesting 240 acres of corn. Peter Pratt, a thirty-five-year-old farmer from Dixon, Illinois, stated, "The farming community will come together to help each other out," and that assisting the beleaguered family would help Shippert to "focus on getting better and getting healthy." Pratt continued, "This is one less thing he has to deal with."[35] Such stories serve to remind all Americans of the many sacrifices Corn Belt families have made to provide for an ever-growing and increasingly urban population.

NOTES

◇

CHAPTER 1

1. Jim Klein, interview by Derek Oden, digital recording, December 15, 2012.

2. Larry Glenn, interview by Derek Oden, digital recording, December 28, 2012.

3. Price V. Fishback and Shawn Everett Kantor, *A Prelude to the Welfare State: The Origins of Workers' Compensation* (Chicago: University of Chicago Press, 2006), 3.

4. Ruth Falck, interview by Derek Oden, digital recording, March 5, 2013.

5. James Wilson, interview by Derek Oden, digital recording, December 30, 2012.

6. "Laverne Jacobson Victim of Tractor Accident," *Roland Record* (Iowa), May 8, 1969, 1.

7. This book seeks to remedy scholars' general neglect of the issue. The lack of a historical survey of the topic is immediately apparent after a brief survey of the literature. Although authors such Andrew Prouty have investigated the history of workplace safety by focusing on a single industry or a specific piece of legislation, farming is rarely mentioned.

8. J. L. Anderson, *Industrializing the Corn Belt: Agriculture, Technology and Environment, 1945–1972* (DeKalb: Northern Illinois University Press, 2009), 5–6, 194. Many other scholars investigating twentieth-century agriculture have used similar terms in describing how farmers' acquisition of new technologies transformed rural landscapes. In 1962, historian Wayne Rasmussen used the term "second revolution" in depicting the dramatic technological and economic changes that had recently occurred; see Wayne D. Rasmussen, "The Impact of Technological Change on American Agriculture, 1862–1962," *The Journal of Economic History* 22 (December 1962): 578–591. Later, John L. Shover made similar observations, concluding that rapid innovation had "turned traditional farming upside down"; see Shover, *First Majority–Last Minority: The Transforming of Rural Life in America* (DeKalb: Northern Illinois University Press, 1976), 5. More recently, historians such as Deborah Fitzgerald and Stephen Stoll have incorporated the term "industrial" and associated terms into depictions of such change; see Deborah Fitzgerald, *Every Farm a Factory: The Industrial Ideal in American Agriculture* (New Haven, Connecticut: Yale University Press, 2003), 3–5; and Stephen Stoll,

The Fruits of National Advantage: Making the Industrial Countryside in California (Berkeley: University of California Press, 1998), 2.

9. Anderson, *Industrializing the Corn Belt,* 5–6. Shover also emphasized that farmers were increasing the scale of their operations; nonetheless, farms during the mid-twentieth century were overwhelmingly owned by individuals, with the result that family farming dominated Corn Belt agricultural production. For a detailed discussion of these economic and demographic issues, see Shover, *First Majority–Last Minority,* 164. For additional perspective on the dominance in agricultural production of family farming operations throughout the early to mid-twentieth century, especially in the Midwest, see Katherine Jellison, *Entitled to Power: Farm Women and Technology, 1913–1963* (Chapel Hill: University of North Carolina Press, 1993), 5.

10. Scholars do not fully agree on what constitutes the American Midwest. In fact, R. Douglas Hurt states, "The Midwest is an amorphous region. Neither scholars nor the public, including those that live in the region, agree on its boundaries or what is included"; quoted in foreword to J. L. Anderson, ed., *The Rural Midwest since World War II* (De Kalb: Northern Illinois University Press, 2014), ix.

11. R. Douglas Hurt, *American Agriculture: A Brief History,* rev. ed. (West Lafayette, Indiana: Purdue University Press, 2002), 338. Katherine Jellison stressed that by the beginning of the First World War corn had become "the major cash crop" in the states of Indiana, Illinois, Iowa, Minnesota, Missouri, Wisconsin, Kansas, Nebraska, and South Dakota; see Jellison, *Entitled to Power,* 5; these states will be the geographical focal points of this discussion of the farm safety issue.

12. John C. Hudson, *Making the Corn Belt: A Geographical History of Middle-Western Agriculture* (Bloomington: Indiana University Press, 1994), 1.

13. Anderson, *Industrializing the Corn Belt,* 190–196, and Stoll, *The Fruits of National Advantage,* 16–24.

14. "Threshing Runs in Galt and Como Are Through for Year," *Sterling Daily Gazette* (Illinois), August 11, 1941, 4.

15. For a more detailed treatment of the increasing emphasis on consumer rights and product safety across a variety of industries, including automobiles, food, and x-rays, from the 1960s onward, see Justin Martin, *Nader: Crusader, Spoiler, Icon* (Cambridge, Massachusetts: Perseus Publishing, 2002), 38–48, 64–74.

16. John Van Ringelestein, interview by Derek Oden, digital recording, December 28, 2012.

17. "The Farmer's Opinion," *Cedar Rapids Gazette* (Iowa), April 20, 1950, 18.

18. Bill Tudor, "Outstanding Farmer Works for Future Successes," *Le Mars Daily Sentinel* (Iowa), March 22, 1966, 3.

19. Dennis Nordin and Roy V. Scott, *From Prairie Farmer to Entrepreneur: The Transformation of Midwestern Agriculture* (Bloomington: Indiana University Press, 2005), 115, 131.

20. Ibid., 139.

21. Ibid., 145.

22. Scholars have been investigating the various issues related to expertise and its influence on American culture and politics for decades. For some useful examples from this lengthy historiography, see Monte A. Calvert, *The Mechanical Engineer in America, 1830–1910* (Baltimore: Johns Hopkins University Press, 1967), 317; Thomas L. Haskell, *The Emergence of Professional Social Science: The American Social Science Association and the Nineteenth Century Crisis of Authority* (Urbana: University of Illinois Press, 1977), 276; and Gerald N. Grob, *Mental Illness and American Society, 1875–1940* (Princeton, New Jersey: Princeton University Press, 1983), 448. For more recent discussions of the influence of expertise in areas as diverse as mental health, auto mechanics, and the military-industrial complex, see Allan V. Horwitz, *Creating Mental Illness* (Chicago: University of Chicago Press, 2002), 315; Kevin L. Borg, *Auto Mechanics: Technology and Expertise in Twentieth Century America* (Baltimore: Johns Hopkins University Press, 2010), 280; Joy Rohde, *Armed with Expertise: The Militarization of Social Research during the Cold War* (Ithaca, New York: Cornell University Press, 2013), 224.

23. Burton J. Bledstein, *The Culture of Professionalism: The Middle Class and the Development of Higher Education in America* (New York: W. W. Norton, 1976), 33–34.

24. Ibid., 37, 84.

25. Ibid., 88.

26. For a discussion of the emergence and early influence of the industrial managerial class, see Robert Kanigel, *The One Best Way: Frederick Winslow Taylor and the Enigma of Efficiency* (New York: Penguin Books, 1997), 233–235, 281–283, 352–353.

27. Danbom, *Born in the Country*, 175.

28. The Midwest newspaper record provides an effective if admittedly incomplete window into the reality of farm accidents during the period prior to this study. For examples see "Shocking Accident," *Belleville News Democrat* (Illinois), August 22, 1902, 1; "Corn Shredder Takes Off Arm," *Aberdeen Daily American* (South Dakota), February 2, 1912, 8; "Skull Smashed in Accident on Farm," *Grand Forks Daily Herald* (North Dakota), October 17, 1915, 3.

29. Claire Swisher, "An Editor Talks about Pitchforks," *Farm Safety Review*, September/October 1943, 7. Swisher edited the *Hammermill Bond*, a publication printed exclusively for the employees of the Hammermill paper company. Another *Farm Safety Review* article on the theme appeared two years later: "Pitch Forks Are Dangerous," *Farm Safety Review*, November/December 1945, 9.

30. Larry Glenn, interview by Derek Oden, digital recording, December 28, 2012.

31. "Tractor Gaining Popularity in the West," *Postville Herald* (Iowa), July 26, 1918, 3.

32. "Minnesota Model for Safety," *Farm Safety Review*, January/February 1946,

7; for a discussion of pioneering work on farm safety in other states, see J. H. Mohler, "150 Lives Saved: The Story of the Famous Kansas Farm Safety Program," *Farm Safety Review,* May/June 1944, 3.

33. "Farm Equipment Institute Convention," *Farm Implement News,* November 3, 1938, 3.

34. *Transactions: National Safety Congress* (Chicago: National Safety Council, 1938), 33.

35. Herb Plambeck, ed., *Iowa Farm Safety in the Twentieth Century: A History of Contributions by Rural Safety Volunteers* (Des Moines: R. E. Hays and Associates, 1983), 133. Norval Wardle also indicated that state-level farm safety activity had begun to emerge by the 1930s. In an unpublished history of the farm safety movement, Wardle notes that states such as Minnesota had small but active state-level safety programs involving such groups as 4-H. See Norval Wardle, "Safe Living for Farm People" (unpublished history of the farm safety movement, 1966), Special Collections, Parks Library, Iowa State University, Ames.

36. U.S. Department of Agriculture, Bureau of Agricultural Economics, *The Prevention of Accidents on Farms and Homes* (Washington, D.C., 1942).

37. Mark Aldrich, *Safety First: Technology, Labor and Business in the Building of American Work Safety, 1870–1939* (Baltimore: Johns Hopkins University Press, 1997), 262–263, 280.

38. Ibid., 11–23, 169, 23; a coupling accident refers to the many injuries or deaths occurring when workers attempted to manually attach rail cars; see Aldrich, *Safety First* 22–23; for an earlier but nonetheless informative discussion of railroad accidents in a specific region of the United States, see Edward Chase Kirkland, *Men, Cities, and Transportation: A Study in New England History, 1820–1900* (Cambridge, Massachusetts: Harvard University Press, 1948), 235–236, 352–361.

39. Aldrich, *Safety First,* 178, 184, 181, 188.

40. Keith Dix, *What's a Miner to Do? The Mechanization of Coal Mining* (Pittsburgh, Pennsylvania: University of Pittsburgh Press, 1988), 101–104; William Graebner, *Coal-Mining Safety in the Progressive Period* (Lexington: University of Kentucky Press, 1976), 15, 1–10.

41. Aldrich, *Safety First,* 216–220, 72, 227, 233–336.

42. Ibid., 80–82, 94, 101.

43. Ibid., 123, 91–92, 129.

44. For a more detailed discussion of the history of chemical regulation prior to the 1970s, see Gino J. Marco, Robert M. Hollingsworth, and Jack R. Plimmer, eds., *Regulation of Agrochemicals: A Driving Force in Their Evolution* (Washington, D.C.: American Chemical Society, 1991), 1–7.

45. U.S. Department of Health, Education, and Welfare, *Facts for Consumers: Pesticide Residues* (Washington, D.C., 1963), 1–2.

46. U.S. Department of Agriculture, Agricultural Research Service, *Behind the Pesticide Label* (Washington, D.C., 1964), 2.

47. U.S. Department of Health, Education, and Welfare, *Facts for Consumers: Pesticide Residues,* 1, 4, 6.

48. Walter T. Reed, "Some Effects of the Miller Law," *Agricultural Chemicals* 10 (September 1955): 38.

49. Marco, Hollingsworth, and Plimmer, eds., *Regulation of Agrochemicals,* 4.

50. *Transactions: National Safety Congress* (Chicago: National Safety Council, 1959), 13; U.S. Department of Agriculture, Agricultural Research Service, *Behind the Pesticide Label,* 3–4.

51. Peter C. English, *Old Paint: A Medical History of Childhood Lead-Paint Poisoning in the United States to 1980* (New Brunswick, New Jersey: Rutgers University Press, 2001), 31–35, 42, 54.

52. Christopher C. Sellers, *Hazards of the Job: From Industrial Disease to Environmental Health Science* (Chapel Hill: University of North Carolina Press, 1997), 227–229.

53. Marilyn Adams with Mary Kay Shanley, *Rhythm of the Seasons: A Journey Beyond Loss* (Marshalltown, Iowa: Sta-Kris, Inc., 1997), 12–13.

54. "Farm Safety For Just Kids: 2003 Annual Report" (Annual Report of Farm Safety For Just Kids, Earlham, Iowa, 2003), 2.

55. Ibid., 6.

CHAPTER 2

1. Teresa Klein, interview by Derek Oden, digital recording, December 15, 2012. Farm children were particularly vulnerable because little legislation protected them. The few legal safeguards that existed applied only to individuals working as farm employees and did not protect the thousands of children who worked on their own families' farms. Thus, parents were ultimately responsible for their children's welfare and for assigning them age-appropriate work duties. For further insights regarding labor laws and farm dangers associated with farm children, see U.S. Department of Labor, Employment Standards Administration, Wage and Labor Division, *Child Labor Requirements in Agriculture Under the Fair Labor Standards Act,* Child Labor Bulletin No. 105 (Washington, D.C., 1984), 1–5; and Harold Heldreth, "Safety for Farm Children," *Farm Safety Review,* March/April 1955, 12.

2. *Transactions: National Safety Congress* (Chicago: National Safety Council, 1954), 24.

3. Maynard H. Coe, "Why a Farm Safety Week?" *Farm Safety Review,* July/August 1946, 14.

4. Maynard H. Coe, "A Family Affair," *Farm Safety Review,* March/April 1959, 1.

5. Rosemary Schaefer, "Safety for Farm Children," *Farm Safety Review,* July/August 1954, 12.

6. "Move to Curtail Accidents with Blasting Caps," *Farm Safety Review*, March/April 1944, 14.

7. Heldreth, "Safety for Farm Children," 10.

8. U.S. Department of Labor, Employment Standards Administration, Wage and Labor Division, *Child Labor Requirements in Agriculture*.

9. Heldreth, "Safety for Farm Children," 12.

10. *Transactions: National Safety Congress* (Chicago: National Safety Council, 1963), 25.

11. Jim Klein, interview by Derek Oden, digital recording, December 15, 2013.

12. Pamela Riney-Kehrberg, *Childhood on the Farm: Work, Play, and Coming of Age in the Midwest* (Lawrence: University Press of Kansas, 2005), 44.

13. Ibid.

14. Ibid., 45.

15. Leroy Hogeland, interview by Derek Oden, digital recording, January 3, 2013.

16. James Wilson, interview by Derek Oden, digital recording, December 30, 2012.

17. Charles and Rose Mary Scott, interview by Derek Oden, digital recording, June 16, 2014.

18. Bernard Stodghill, interview by Derek Oden, digital recording, December 29, 2012.

19. "Could This Be Your Child?" *Farm Safety Review*, September 1955, 8.

20. Schaefer, "Safety for Farm Children," 13.

21. Ibid.

22. Heldreth, "Safety for Farm Children," 11.

23. Ibid.

24. Lawrence Schmitz, interview by Derek Oden, digital recording, December 28, 2012.

25. Leroy Hogeland, interview by Derek Oden, digital recording, January 3, 2013.

26. *Transactions: National Safety Congress* (Chicago: National Safety Council, 1962), 2.

27. Craig Fausch, interview by Derek Oden, digital recording, January 3, 2013.

28. Don Rimathe, interview by Derek Oden, digital recording, January 3, 2013.

29. Jim Klein, interview by Derek Oden, digital recording, December 15, 2013.

30. James Wilson, interview by Derek Oden, digital recording, December 30, 2013.

31. Harold Heldreth, "Providing Safe Farms for the Older People in Agriculture," *Farm Safety Review*, January/February 1953, 12–14.

32. Arthur W. Turner, "Let's Be Sensible about Farm Safety," *Farm Safety Review*, November/December 1946, 9.

33. James Wilson, interview by Derek Oden, digital recording, December 30, 2012.

34. Bernard Stodghill, interview by Derek Oden, digital recording, December 29, 2012.

35. D. B Pointer, "Part-Time Farming Way of Life," *Jefferson City Post-Tribune* (Missouri), April 10, 1969, 15. The issue of part-time farmers appeared occasionally in midwestern newspapers throughout the era; see W. D. Davison, "Country Squire Faces Challenges," *Waterloo Sunday Courier* (Iowa), January 20, 1960, 20; Rodney Hayden, "Part-Time Farming Costly," *The Advocate* (Newark, Ohio), September 10, 1975, 30.

36. Bernard Stodghill, interview by Derek Oden, digital recording, December 29, 2012.

37. Marvin J. Nichol, "How's Your Horse Sense?" *Farm Safety Review*, May/June 1945, 8.

38. Rex Peterson, interview by Derek Oden, digital recording, January 1, 2013.

39. Perry Middlesworth, interview by Derek Oden, digital recording, December 29, 2012.

40. Jim Klein, interview by Derek Oden, digital recording, December 15, 2012.

41. Leroy Hogeland, interview by Derek Oden, digital recording, January 3, 2013.

42. Rex Peterson, interview by Derek Oden, digital recording, January 1, 2013.

43. Harold E. Heldreth, "Safe Practices Can Mean Bigger Profits," *Farm Safety Review*, August 1956, 8.

44. E. S. Bayard, "Put the Bull in His Place," *Farm Safety Review*, September/October 1943.

45. Randall C. Swanson, "The Dairy Bull—Dr. Jekyll and Mr. Hyde," *Hoard's Dairyman*, November 1944, 598.

46. John Van Ringelestein, interview by Derek Oden, digital recording, December 28, 2013.

47. Rex Peterson, interview by Derek Oden, digital recording, January 1, 2013. For additional perspective on the potential dangers of seemingly docile bulls and additional animal accident stories, see "Two Lucky Accidents," *Farm Safety Review*, May/June 1943, 3; and Swanson, "The Dairy Bull—Dr. Jekyll and Mr. Hyde," 598.

48. The bull pen was one of several suggestions for avoiding bull-related accidents. Other suggestions included not handling bulls in fields and other unprotected spaces and not using ropes to "lead" bulls; see Bayard, "Put the Bull in His Place," 9; and Swanson, "The Dairy Bull—Dr. Jekyll and Mr. Hyde," 598.

49. C. L. Hamilton, "The Safety Bull Pen," *Farm Safety Review*, November/December 1944, 7; and Swanson, "The Dairy Bull—Dr. Jekyll and Mr. Hyde," 598.

50. Pat Larkin, interview by Derek Oden, digital recording, December 29, 2012.

51. Heldreth, "Safe Practices Can Mean Bigger Profits," 9.

52. "Animal Disease and Human Health," *Farm Safety Review,* March/April 1965, 12; *Transactions: National Safety Congress* (Chicago: National Safety Council, 1953), 35.

53. *Transactions: National Safety Congress* (Chicago: National Safety Council, 1953), 34.

54. "Animal Disease and Human Health," 15.

55. Burton Williams, "Going to Blazes," *Farm Safety Review,* May/June 1959, 8–9; A. M. Sowder, "Clean for Safety's Sake," *Farm Safety Review,* March/April, 1949, 4; "Safe Cans," *Farm Safety Review,* May/June 1953, 9.

56. *Transactions: National Safety Congress* (Chicago: National Safety Council, 1954), 24; "Fires: Why They Happen," *Farm Safety Review,* July/August 1962, 9.

57. David McFarland, "Eight Steps to Combine Safety," *Farm Safety Review,* May/June 1965, 4; Joseph Gerling, *Extension Facts: Safe Combine Operation* (Stillwater: Oklahoma State University, 1974), 4. Farm safety experts provided additional suggestions for avoiding a machinery-related fire, including never smoking while pumping fuel and checking consistently for gas or diesel leaks. These specific suggestions and others were collected in the National Safety Council's *Bulletin: Grain Harvest Safety* (Chicago: National Safety Council, 1979).

58. Robert Westpfahl, "Another Farmer's View of Safety" (speech given at meeting of the National Institute of Farm Safety, Rock Island, Illinois, June 17, 1981), Norval Wardle Collection, Iowa State University Archives, Iowa State University, Ames.

59. Wesley Longnecker, interview by Derek Oden, digital recording, January 2, 2013.

60. Rex Peterson, interview by Derek Oden, digital recording, January 1, 2013.

61. James Wilson, interview by Derek Oden, digital recording, December 30, 2012.

62. Hurt, *American Agriculture: A Brief History,* rev. ed., 326; for both a more complete discussion of the rate of dissemination and effect of electricity in the countryside, see Bruce L. Gardner, *American Agriculture in the Twentieth Century: How It Flourished and What It Cost* (Cambridge, Massachusetts: Harvard University Press, 2002), 15; *Transactions: National Safety Congress* (Chicago: National Safety Council, 1962), 22.

63. *Transactions: National Safety Congress* (Chicago: National Safety Council, 1964), 22.

64. Myrtle Fahsbender, "Light: For Greater Farm Safety," *Farm Safety Review,* September/October 1944, 7; Earl L. Arnold, "Lighting for Safety," *Farm Safety Review,* January/February 1952, 6.

65. Safety advocates contended that electric lights were safer than earlier light sources; for example, Burton Williams, a farm safety advocate, stated, "Electric lights reduce the farm fire hazards because they replace hazardous lamps and candles" (from Williams, "Going to Blazes," 9).

66. *Transactions: National Safety Congress* (Chicago: National Safety Council,

1962), 22; William Fletcher, *Rural Accident Bulletin: Electrical Safety on the Farm and Ranch* (Chicago: National Safety Council, 1975), 1.

67. William Peterson, "Farmstead Wiring: A Production Tool," *Farm Safety Review,* January 1956, 1; and William H. Knight and Owen K. Brown, "Wire Your Farm for Tomorrow's Load," *Farm Safety Review,* July/August 1949, 13.

68. *Transactions: National Safety Congress* (Chicago: National Safety Council, 1962), 22–23.

69. Peterson, "Farmstead Wiring," 1; and Jack Burke, "Farmstead Fix-up," *Farm Safety Review,* May/June 1968, 5.

70. Fletcher, *Rural Accident Bulletin: Electrical Safety on the Farm and Ranch,* 2.

71. *Transactions: National Safety Congress* (Chicago: National Safety Council, 1952), 20–21; "Heat Lamp Safety," *Farm Safety Review,* July/August 1954, 20; C. L. Hamilton, "Electric Fence Problems," *Farm Safety Review,* November/December 1953, 6; C. L. Hamilton, "Is Your Electric Fence Safe?" *Farm Safety Review,* March/April 1945, 7.

72. Teresa Klein, interview by Derek Oden, digital recording, March 15, 2013.

73. Fletcher, *Rural Accident Bulletin: Electrical Safety on the Farm and Ranch,* 2–3.

74. "It's the Current that Kills: Electricity Can Be Deadly—When Safety Is Neglected," *Farm Safety Review,* November/December 1949, 10.

75. C. L. Hamilton, "Shocking Conditions," *Farm Safety Review,* November/December 1949, 8–9.

76. "Trained Electrician Should Do All Wiring," *Better Iowa,* July 23, 1945, 2; and *Farm Safety Review,* May/June 1968, 5.

77. Burke, "Farmstead Fix-up," 5.

78. H. H. Beaty, "Electricity Can Be Fire Hazard," *Better Iowa,* 24 July 1944, 2; "Trained Electrician Should Do All Wiring," 2.

79. John Van Ringelestein, interview by Derek Oden, digital recording, December 28, 2012.

80. Jim and Teresa Klein, interview by Derek Oden, digital recording, December 15, 2012.

81. Leroy Hogeland, interview by Derek Oden, digital recording, January 3, 2013.

82. G. E. Henderson, "Explaining Electrical Safety," *Farm Safety Review,* November/December 1952, 4–5.

83. "Beware: It's Lightning Season," *Farm Safety Review,* May/June 1960, 11. Farm families had long known the dangers posed by lightning strikes. For an account of the tragic death by lightning of one Minnesota farmer in the late nineteenth century, see Steven R. Hoffbeck, *The Haymakers: A Chronicle of Five Farm Families* (St. Paul: Minnesota Historical Society Press), 79.

84. Ruth Falck, interview by Derek Oden, digital recording, March 15, 2013.

85. Pat Larkin, interview by Derek Oden, digital recording, December 2, 2012.

86. Jim Klein, interview by Derek Oden, digital recording, December 15, 2012.

87. "Recreation: Facts," *Farm Safety Review,* November/December 1959, 10; "Farm Pond and Irrigation Safety," *Farm Safety Review,* March/April 1960, 2; Jack Burke, "How to Make Your Farm Pond Safe," *Farm Safety Review,* July/August 1966, 11.

CHAPTER 3

1. R. Douglas Hurt, *American Farm Tools: From Hand-Power to Steam-Power* (Manhattan, Kansas: Sunflower University Press, 1982), 4–5.

2. Ibid., 4–5, 112.

3. Willard W. Cochrane, *The Development of American Agriculture: A Historical Analysis* (Minneapolis: Development of American Agriculture, 1993), 125.

4. Anderson, *Industrializing the Corn Belt,* 8.

5. Gilbert Fite, *American Farmers: The New Minority* (Bloomington: Indiana University Press, 1981), 66.

6. Lawrence Schmitz, interview by Derek Oden, digital recording, December 28, 2012.

7. Don Paarlberg and Phillip Paarlberg, *The Agricultural Revolution of the Twentieth Century* (Ames: Iowa State University Press, 2000), 24.

8. O. C. Hulett, "For 86 Years Ballacks Have Used Horses to Farm Their 240 Acres," *Racine Journal Times* (Wisconsin), May 6, 1956, 4b. For more on the decline of horse-powered farming, see F. A. Hankins, "Word from the Country, Hankins Would Like More Horse Stories," *Cedar Rapids Gazette* (Iowa), July 21, 1963, 11b.

9. Ibid., 140.

10. Anderson, *Industrializing the Corn Belt,* 165–166.

11. Charles J. Scranton, "Safety and the Mechanical Corn Picker," *Agricultural Engineering* 33, no. 3 (1952): 140.

12. Mark Zimmerman, "Scanning the Self-Propelled Combines," *Implement and Tractor,* June 21, 1965, 20.

13. Anderson, *Industrializing the Corn Belt,* 165–166.

14. Benson J. Lamp and Kenneth A. Harkness, "Recommendations and Improved Design of Portable Elevators" (paper presented at the annual meeting of the American Society of Agricultural Engineers, Chicago, December 15–18, 1959), 1. For more on the labor-saving benefits of elevators and augers, see William Fletcher, "Avoiding Elevator and Auger Accidents," *Farm Safety Review,* November/December 1969, 5.

15. T. David McFarland and William J. Fletcher, *Analysis of Portable Farm Elevator and Auger Accidents to Determine Corrective Measures,* Extension Pamphlet (Chicago, 1969), 3, Norval Wardle Collection, Special Collections, Iowa State University Library, Ames. Keith Roe, an authority on the history of grain storage

practices, indicated that although elevators had been in existence prior to the Great Depression, it was not until the mid-1930s that farmers possessed them in significant numbers; see Keith E. Roe, *Corncribs: In History, Folklife, and Architecture* (Ames: Iowa State University Press, 1988), 41.

16. National Safety Council, Farm Department, *Rural Accident Prevention Bulletin: Wagon Safety* (Chicago: National Safety Council, 1978), 1–2.

17. Gardner, *American Agriculture in the Twentieth Century*, 13. Gardner's book also provides a brief but useful summary of major mechanical innovations that were instrumental in farming (9).

18. Westpfahl, "Another Farmer's View of Safety," 1.

19. Lawrence Schmitz, interview by Derek Oden, digital recording, December 28, 2012.

20. *Transactions: National Safety Congress* (Chicago: National Safety Council, 1953), 32.

21. James Wilson, interview by Derek Oden, digital recording, January 3, 2013.

22. "Practice Safety with Machinery," *Hoard's Dairyman*, August 10, 1943, 466; "Farm Accidents Are Made," *Hoard's Dairyman*, July 20, 1943, 412.

23. Frank J. Zink, "Farm Equipment Can Be Operated Safely" (paper presented at the Nebraska Conference on Home and Farm Safety, Lincoln, Nebraska, October 2, 1942), 5. Zink also revealed that the war had only exacerbated the problem, indicating that millions of inexperienced laborers were entering the agricultural workforce with insufficient training, particularly large numbers of women and children who were replacing the many males serving in the armed services. For a detailed discussion, see Zink, "Farm Equipment Can Be Operated Safely," 6.

24. Jack Burke, "The Lonesome Plowman," *Farm Safety Review*, March/April 1968, 14.

25. Maynard Coe, "Why Farm Safety Week?" *Farm Safety Review*, July/August 1946, 14.

26. Charlie Nettles, "Farmers Tell Machinery Gripes," *Des Moines Sunday Register* (Iowa), September 27, 1970, sec. F, 2.

27. "Practice Safety with Machinery," 466; Scranton, "Safety and the Mechanical Corn Picker," 140.

28. Pat Larkin, interview by Derek Oden, digital recording, December 29, 2012.

29. *Transactions: National Safety Congress* (Chicago: National Safety Council, 1962), 27; National Safety Council, Farm Department, *Rural Accident Bulletin: Wagon Safety*, 3.

30. Perry Middlesworth, interview by Derek Oden, digital recording, December 29, 2012.

31. Daniel Kitchen, "Stop That Picker . . . Before You Leave the Tractor Seat," *Farm Safety Review*, September 1955, 6.

32. C. L. Hamilton, "Adjust Corn Picker: For Safety and Efficiency," *Farm Safety*

Review, September/October 1952, 8–9; and "Pick More Corn Safely," *Farm Safety Review*, September/October 1958, 10.

33. Joseph Gerling, *Extension Facts: Safe Combine Operation* (Stillwater: Oklahoma State University, 1974), 2.

34. Farmers faced a variety of other hazards such as fire and spinning cranks for raising elevators or starting tractors; for examples, see W. E. Stuckey, B. J. Lamp, and K. A. Harkness, *Bulletin 367: The Portable Farm Elevator: Make It a Helper Only* (Columbus: Ohio State University, Agricultural Extension Service, 1962), 4; Lamp and Harkness, "Recommendations for Improved Design of Portable Farm Elevators" (paper presented at the annual meeting of the American Society of Agricultural Engineers, Chicago, December 15–18, 1959), 3.

35. Iowa State University, Cooperative Extension Service, *Fatal Accidents of Iowa Farm People, 1947–1971* (Ames, Iowa: Cooperative Extension Service, 1971), 8–9.

36. Lawrence Schmitz, interview by Derek Oden, digital recording, December 28, 2012.

37. Hurt, *American Agriculture: A Brief History*, rev. ed., 337.

38. Cochrane, The Development of American Agriculture, 108–109. In 1962, FFA delegates from Ellsworth, Kansas, speaking at the National Safety Council Congress alerted attendees to the magnitude of the problem in their state. They indicated that "[d]uring 1961 there were twenty-seven persons who were careless in Kansas, and were killed. Tractors caused over half of the forty-seven farm fatalities in our state last year"; see *Transactions: National Safety Congress* (Chicago: National Safety Council, 1962), 27.

39. Larry Glenn, interview by Derek Oden, digital recording, December 28, 2012.

40. Jim Klein, interview by Derek Oden, digital recording, December 15, 2012.

41. "Next Time . . . This Might Be You!" *Wallaces Farmer and Iowa Homestead*, June 7, 1952, 8.

42. Rex Peterson, interview by Derek Oden, digital recording, January 1, 2013.

43. Don Rimathe, interview by Derek Oden, digital recording, January 3, 2013.

44. National Safety Council, *How to Get Your Wagon Rolling Safely* (Chicago: National Safety Council, 1968), 2.

45. National Safety Council, Farm Department, *Rural Accident Prevention Bulletin: Wagon Safety* (Chicago: National Safety Council, 1978), 1.

46. Ibid. National Safety Council, *How to Get Your Wagon Rolling Safely*, 3.

47. Jim Klein, interview by Derek Oden, digital recording, December 15, 2012.

48. Sam Brungardt, "Safe Handling of Big Bales," *Implement and Tractor*, March 1976, 12.

49. Dr. Richard L. Jepsen, a farm safety specialist from Kansas State University, believed the "three-point-hitch bale handling device to be safer than the modified front end loaders." He also thought farmers would persist in using adapted

front-end loaders even though they offered less stability. Rollin Schnieder, a farm safety expert employed at the University of Nebraska, echoed Jepsen's opinions. Schnieder stated "that some of the front end loaders which are modified in farm shops may be poorly engineered and put excess stress on a tractor's hydraulic system." See Brungardt, "Safe Handling of Big Bales," 12.

50. Lawrence Schmitz, interview by Derek Oden, digital recording, December 28, 2012. Children encountered dangers while playing with other farm equipment as well, such as elevators. Farm youth would sometimes climb elevators and be seriously injured when the devices tipped over or collapsed; see Stuckey, Lamp, and Harkness, *Bulletin 367: The Portable Farm Elevator: Make It a Helper Only*, 7.

51. Perry Middlesworth, interview by Derek Oden, digital recording, December 29, 2012. Despite such accident problems, engineers were making progress by adding features such as "safety tracks" to elevators to provide greater stability. Unfortunately, such accident prevention devices were not on all new models in the late 1960s. For more perspective on the many aspects of elevator safety for both children and adults, see Stuckey, Lamp, and Harkness, *Bulletin 367: The Portable Farm Elevator: Make It a Helper Only;* McFarland and Fletcher, *Analysis of Portable Farm Elevator and Auger Accidents to Determine Corrective Measures*, 23–24, 33–34.

52. *Transactions: National Safety Congress* (Chicago: National Safety Council, 1962), 27; E. W. Tanquary, "Safeguarding Power-Driven Farm Machinery" (paper prepared for the members of the Farm Equipment Institute's Advisory Engineering Committee, Chicago, 1950), Norval Wardle Collection, Special Collections, Iowa State University Library, Ames. Despite Tanquary's observations, many farmers appear to have taken the use of guards seriously; for instance, Bernard Stodghill commented that he always kept such devices attached to his equipment, since he was fully aware of the dangers; see Bernard Stodghill, interview by Derek Oden, digital recording, December 29, 2012.

53. McFarland and Fletcher, *Analysis of Portable Farm Elevator and Auger Accidents to Determine Corrective Measures*, 33.

54. Norval Wardle, "Safety Controls and Shielding for Augers" (paper presented at the Annual Meeting of the American Society of Agricultural Engineers, Hot Springs, Arkansas, June 28, 1972), 3–4. Found in Norval Wardle Special Collections, Parks Library, Iowa State University, Ames, Iowa.

55. Wesley Longnecker, interview by Derek Oden, digital recording, January 2, 2013.

56. Rex Peterson, interview by Derek Oden, digital recording, January 1, 2013.

57. Perry Middlesworth, interview by Derek Oden, digital recording, December 29, 2012.

58. "Winter Increases Peril of PTO's for Farmers," *Globe Gazette* (Mason City, Iowa), January 14, 1966, 3.

59. "Pick More Corn-Safely," 8–9; Ray Franklin, "Don't Lose an Arm," *Farm Safety Review*, September/October 18, 1952, 8.

60. Kitchen, "Stop That Picker," 5.

61. Leroy Hogeland, interview by Derek Oden, digital recording, January 3, 2013. Farm accident studies conducted at the time confirm the prevalence of such accidents. In the 1940s, the Illinois Agricultural Association conducted an investigation that revealed that Illinois farmers experienced approximately two hundred corn-picking accidents per year in the early to mid 1940s; see Hamilton, "Adjust Corn Picker," 8. A 1951 study in Iowa catalogued the grisly toll: "299 fingers, 32 thumbs and 32 hands"; see Franklin, "Don't Lose an Arm," 8, and *Wallaces Farmer and Iowa Homestead*, 8.

62. Rex Peterson, interview by Derek Oden, digital recording, January 1, 2013.

63. Bernard Stodghill, interview by Derek Oden, digital recording, December 29, 2012.

64. James Wilson, interview by Derek Oden, digital recording, December 30, 2012.

CHAPTER 4

1. James Wilson, interview by Derek Oden, digital recording, December 30, 2012.

2. Leroy Hogeland, interview by Derek Oden, digital recording, January 3, 2013.

3. John Van Ringelestein, interview by Derek Oden, digital recording, December 28, 2012.

4. Scholars such as Thomas Dunlap echo the comments of these Iowa farmers, indicating that many in the agricultural community viewed chemical use similarly; see Thomas Dunlap, *DDT: Scientists, Citizens, and Public Policy* (Princeton, New Jersey: Princeton University Press, 1981), 30–35.

5. David B. Danbom, *Born in the Country: A History of Rural America* (Baltimore: Johns Hopkins University Press, 1995), 237.

6. *Transactions: National Safety Congress* (Chicago: National Safety Council, 1959), 13. See also Mary K. Farinholt, *The New Masked Man in Agriculture* (Cleveland, Ohio: National Consumers Committee for Research and Education, Inc.), 7, 1. For more discussion of the tremendous changes associated with the introduction of a diverse range of agricultural chemicals immediately following the Second World War, see U.S. Department of Health, Education, and Welfare, *Facts for Consumers: Pesticide Residues* (Washington, D.C., October 1963), 1. For a scholarly discussion of the use of agricultural chemicals in the era prior to the era covered in this book, see Dunlap, *DDT*, 29–43.

7. *Transactions* (1959), 1; "Here's Why You Need a Pesticide Safety Program," *Farm Safety Review*, November/December 1965, 3.

8. U.S. Department of Agriculture, Agricultural Research Service, *Safe Use of*

Agricultural and Household Pesticides (Washington, D.C., 1967), i; *Transactions* (1959), 13.

9. Farinholt, *New Masked Man in Agriculture*, 31. Historian David Danbom has also commented on this point stating, "The very effectiveness of DDT proved to be a curse. For one thing, it led to a de-emphasis of research on biological pest control, a more environmentally benign method." See Danbom, *Born in the Country*, 237.

10. An investigation of anhydrous ammonia is particularly germane to this discussion not only because its use became so prevalent during the period, but also because extension agents, farm safety experts, and others frequently cited it as a substance involved in accidents.

11. Scott R. Helmers, W. H. McConnell, and L. W. Knapp, "Ammonia Mishaps in Agriculture," *Agricultural Chemicals* 52, no. 3 (March 1971): 118.

12. Gardner, *American Agriculture in the Twentieth Century*, 23.

13. Anderson, *Industrializing the Corn Belt*, 190–196.

14. Pete Daniel, *Pesticides and Health in the Post–World War II South* (Baton Rouge: Louisiana State University Press, 2005), 7.

15. Lawrence Schmitz, interview by Derek Oden, digital recording, December 28, 2012.

16. Bernard Stodghill, interview by Derek Oden, digital recording, December 29, 2012.

17. N. B. Akesson, J. B. Bailey, W. F. Serat, and W. E. Yates, "Health Hazards to Workers from Application of Pesticides" (paper presented at the Annual Meeting of Agricultural Engineers, Stillwater, Oklahoma, June 23–26, 1974), 1. For a detailed discussion of the largely ineffective and lax regulatory structure following World War II, see Daniel, *Toxic Drift*, 1–209.

18. Danbom, *Born in the Country*, 237.

19. "Spray Safe," *Farm Safety Review*, March/April 1946, 4; Frank Princi, "Toxicology and Hazard Record of the New Pesticides," *Agricultural Chemicals* 7, no. 1 (January 1952): 47.

20. Farinholt, *New Masked Man in Agriculture*, 3.

21. *Transactions: National Safety Congress* (Chicago: National Safety Council, 1962), 4. Pete Daniel also stressed the poor statistical understanding of the issue; see Daniel, *Toxic Drift*, 71–72.

22. William J. Fletcher, "Pesticide Worker Accidents" (paper presented at the National Conference on Protective Clothing and Safety Equipment for Pesticide Workers, Center for Disease Control, Atlanta, Georgia, May 1–2, 1972), 1, 4. For a detailed discussion of the development and scope of this study in the 1950s and 1960s, see Howard W. Mitchell, "Pesticides and Other Agricultural Chemicals as a Public Health Problem," *American Journal of Public Health and the Nation's Health* 55, no. 7 (July 1965): 10–15.

23. Akesson, Bailey, Serat, and Yates, "Health Hazards to Workers from Application of Pesticides," 1.

24. National Safety Council, *Rural Accident Prevention Bulletin: Safe Use of Pesticides* (Chicago: National Safety Council, 1976), 1.

25. "Here's Why You Need a Pesticide Safety Program," *Farm Safety Review*, November/December 1965, 3. A number of writers including, most recently, Pete Daniel have seriously questioned the view that agricultural chemical manufacturers had to undergo an intensive registration process prior to gaining acceptance by government regulatory agencies; see Daniel, *Toxic Drift*, 7.

26. Harold Heldreth, "Read the Label—And Poison Only the Pests," *Farm Safety Review*, June 1955, 11.

27. Farinholt, *New Masked Man in Agriculture*, 6.

28. Ibid., 16.

29. William J. Fletcher, "Mix Well—With Safety," *Farm Safety Review*, March/April 1969, 11.

30. Bernard Stodghill, interview by Derek Oden, digital recording, December 29, 2012.

31. Perry Middlesworth, interview by Derek Oden, digital recording, December 29, 2012.

32. Pat Larkin, interview by Derek Oden, digital recording, December 29, 2012.

33. James Wilson, interview by Derek Oden, digital recording, December 30, 2012.

34. Jim Klein, interview by Derek Oden, digital recording, December 15, 2012.

35. David E. Baker and David B. Smith, *First Aid for Pesticide Poisoning* (Columbia: University of Missouri, Columbia Extension Division, 1976), 1.

36. National Safety Council, *Rural Accident Prevention Bulletin: Safe Use of Pesticides*, 1.

37. Ruth Falck, interview by Derek Oden, digital recording, March 15, 2013. For more perspective on the problem of chemical containers and children in agricultural settings, see U.S. Department of Agriculture, Agricultural Research Service, *Safe Use of Agricultural and Household Pesticides*, 2; Donald L. Miller, *Farm Safety Review*, March/April 1958, 13; *Transactions: National Safety Congress* (Chicago: National Safety Council, 1962), 7; Harold Heldreth, "Insecticides Are Poisonous!" *Farm Safety Review*, March/April 1951, 10.

38. "Insecticide Dangers . . . Carelessness Can Cause Death," *Wallaces Farmer*, March 12, 1983, 84.

39. F. H. Leavitt, "Use and Marketing of Agricultural Ammonia," *Agricultural Chemicals* 8, no. 6 (June 1953): 36; "New Business Develops . . . Anhydrous Ammonia Application," *Agricultural Chemicals* 8, no. 2 (February 1953): 46.

40. U.S. Department of Labor, Occupational Safety and Health Administration, U.S. Department of Agriculture Committee on Safety, *Safe Use of Anhydrous Ammonia* (Washington, D.C., 1972), 2, 4.

41. Don Rimathe, interview by Derek Oden, digital recording, January 3, 2013.

42. Perry Middlesworth, interview by Derek Oden, digital recording, December 29, 2012. John Van Ringelestein remembered once watching his neighbor running from a cloud of anhydrous that had escaped during the application process, John Van Ringelestein, interview by Derek Oden, digital recording, December 28, 2012.

43. James Wilson, interview by Derek Oden, digital recording, December 28, 2012.

44. J. B. Liljedahl, Robert C. Rund, and F. R. Willsey, "A Survey of Anhydrous Ammonia Accidents in Indiana" (report prepared for the National Safety Council regarding anhydrous ammonia victims in Indiana in 1971 and 1972), Norval Wardle Collection, Iowa State University Library, Ames, 1. William J. Fletcher, an agricultural engineer, also argued that rushing to complete work often contributed to anhydrous accidents; see William J. Fletcher, "Analysis of Anhydrous Ammonia Accidents to Determine Corrective Measures" (report supported by an NSC grant, completed May 1974), Norval Wardle Collection, Iowa State University Library, Ames, 13.

45. Liljedahl, Rund, and Willsey, "A Survey of Anhydrous Ammonia Accidents in Indiana," 3; Fletcher, "Analysis of Anhydrous Ammonia Accidents to Determine Corrective Measures," 12.

46. Fletcher, "Analysis of Anhydrous Ammonia Accidents to Determine Corrective Measures," 5.

47. Liljedahl, Rund, and Willsey, "A Survey of Anhydrous Ammonia Accidents in Indiana," 1.

48. Fletcher, "Analysis of Anhydrous Ammonia Accidents to Determine Corrective Measures," 8–10.

49. Arlo Jacobsen, "Operation Peach Warns Farmers of Ammonia Dangers," *Des Moines Register*, April 9, 1972, 1-F. In the same year, the author of an OSHA pamphlet supported Jacobsen's comments, emphasizing that "worn hoses, loose connections, and other defects can cause accidents"; see U.S. Department of Labor, Occupational Safety and Health Administration, U.S. Department of Agriculture Committee on Safety, *Safe Use of Anhydrous Ammonia*, 2.

50. Fletcher, "Analysis of Anhydrous Ammonia Accidents to Determine Corrective Measures," 11; "First Aid Rules: In Case of Ammonia Burns or Injury," *Farm Safety Review*, March 1956, 118. The advice to seek professional assistance appears to have been given for a growing number of issues on the farm, reflecting the increasing technical demands placed on the era's farmers.

51. Helmers, McConnell, and Knapp, "Ammonia Mishaps in Agriculture," 118; Jacobsen, "Operation Peach Warns Farmers of Ammonia Dangers," 5.

52. *Transactions: National Safety Congress* (Chicago: National Safety Council, 1962), 5. For a similar point of view, see J. N. Roney, "Insecticides in the Southwest," *Farm Safety Review*, July/August 1960, 5.

53. *Safe Use of Agricultural and Household Pesticides*, 4.

54. E. P. Sylvester, Harold Gunderson, and Gayle L. Worf, *Suggestions for Safety Spraying: Herbicides, Insecticides, and Fungicides* (Ames: Iowa State University, Cooperative Extension Service, 1963), 1.

55. Leroy Hogeland, interview by Derek Oden, digital recording, January 3, 2013.

56. James Wilson, interview by Derek Oden, digital recording, December 30, 2012.

57. Lawrence Schmitz, interview by Derek Oden, digital recording, December 28, 2012.

58. Don Rimathe, interview by Derek Oden, digital recording, December 28, 2012.

59. *Transactions: National Safety Congress* (Chicago: National Safety Council, 1962), 7.

60. Rex Peterson, interview by Derek Oden, digital recording, January 1, 2013.

61. Farinholt, *New Masked Man in Agriculture*, 15.

62. "Don't Be a Hay Field Oldfield . . . and It's Just as Dangerous to Spray Poisonous Insecticides Without Your M.S.A. Farm Spray Respirator," Mine Safety Appliance Company advertisement, *Farm Safety Review*, June 1955, 15.

63. R. A. Fulton, Floyd F. Smith, and Ruth Bugsby, *Respiratory Devices for Protection against Certain Pesticides* (Washington, D.C.: U.S. Department of Agriculture, Agricultural Research Service, June 1964), 1, Norval Wardle Collection, Special Collections, Iowa State University, Ames.

64. National Safety Council, *Rural Accident Prevention Bulletin: Safe Use of Pesticides*, 2.

65. Farinholt, *New Masked Man in Agriculture*, 15–17.

66. Sylvester, Gunderson, and Worf, *Suggestions for Safety Spraying*, 1.

67. Farinholt, *New Masked Man in Agriculture*, 8–12; Roney, "Insecticides in the Southwest," 5. In 1952, Frank Princi also emphasized the hazards associated with organic phosphorous compounds: "Perhaps the most toxic of the new pesticides are that group known as the organic phosphorous compounds. In common with many other insecticidal materials, these substances may be absorbed through the skin, respiratory tract, conjunctive and gastro-intestinal tract." See Princi, "Toxicology and Hazard Record of Newer Pesticides," 46–47; *Transactions: National Safety Congress* (Chicago: National Safety Council, 1955), 15.

68. Daniel, *Toxic Drift*, 138–139, 136.

69. Publication produced cooperatively by National Agricultural Chemicals Association, National Safety Council, U.S. Department of Agriculture, U.S. Department of Health, Education, and Welfare, National Vocational Agricultural Teachers' Association, Manufacturing Chemists' Association, Inc., *Chemicals in Agriculture: Be Safe! Use Them Properly* (Washington, D.C., January 1964), 2.

70. Burke, *Rural Accident Prevention Bulletin: Safe Use of Pesticides*, 2.

71. Farinholt, *New Masked Man in Agriculture*, 6.

72. Jim Klein, interview by Derek Oden, digital recording, December 15, 2012.

73. Perry Middlesworth, interview by Derek Oden, digital recording, December 29, 2012.

74. Bernard Stodghill, interview by Derek Oden, digital recording, December 29, 2012.

CHAPTER 5

1. Plambeck, ed., *Iowa Farm Safety in the Twentieth Century*, 32.

2. Robert T. Legge, "Occupational Hazards in the Agricultural Industries," *American Journal of Public Health and the Nation's Health* 25, no. 4 (April 1935): 457.

3. Mark Aldrich, *Safety First: Technology, Labor and Business in the Building of American Work Safety, 1870–1939* (Baltimore: Johns Hopkins University Press, 1997), 262. Although both Mark Aldrich and contemporary observers believed that the mid-twentieth century witnessed an increase in farm accidents, little farm accident data from the decades prior to the 1940s is available to support the contention. Allan B. Kline, president of the American Farm Bureau Federation, stated that "the increased mechanization of the farm operations, plus the speed at which we operate, has vastly increased the hazards to which all farm people are exposed every day"; see "Agricultural Leaders Endorse Farm Safety Week's Objective," *Farm Safety Review*, July/August 1948, 3.

4. Theodore Porter, *The Rise of Statistical Thinking, 1820–1900* (Princeton, New Jersey: Princeton University Press, 1986), 104. For additional perspective regarding the impact of statistical thinking, especially by the life insurance industry, see Dan Bouk, *How Our Days Became Numbered: Risk and the Rise of Statistical Individual* (Chicago: University of Chicago Press, 2015), xi–xii. Historian John C. Burnham emphasized the importance of the maturation of statistical thinking in raising awareness of industrial accidents in late nineteenth-century Europe; see John C. Burnham, *Accident Prone: A History of Technology, Psychology, and Misfits of the Machine Age* (Chicago: University of Chicago Press, 2009), 9.

5. The farm safety movement included not only youth organizations, farm safety committees, and farm safety experts, but also many other historical actors including the Farm Bureau. For an early example of the safety activities of a Farm Bureau chapter, in this case one located in Olmstead County, Minnesota, see "There's a Prize for Everyone," *Farm Safety Review*, July/August 1945, 5.

6. *Transactions: National Safety Council* (Chicago: National Safety Council, 1913), 69, 72; *Transactions: National Safety Council* (Chicago: National Safety Council, 1925), 6. For a discussion of the many work hazards in the nation's

manufacturing sectors during the late nineteenth and early twentieth centuries, see Aldrich, *Safety First*, 79–89.

7. *Transactions: National Safety Council* (Chicago: National Safety Council, 1913), 69. In 1944, Ned H. Dearborn, NSC vice president, stated that a Chicago headquarters' staff member "compiles information, prepares and publishes educational materials, and stimulates and acts as a clearinghouse for the work of state and community councils and outside agencies"; see Ned H. Dearborn, "What Is the National Safety Council?" *Farm Safety Review*, September/October 1944, 3.

8. Aldrich, *Safety First*, 106; *Transactions: National Safety Council* (Chicago: National Safety Council, 1914), 18; *Transactions: National Safety Council* (Chicago: National Safety Council, 1918), 68; *Transactions: National Safety Council* (Chicago: National Safety Council,1919), 212.

9. Aldrich, *Safety First*, 109–110.

10. Ibid., 108–110, 114, 103; *Transactions: National Safety Council* (Chicago: National Safety Council, 1938), 9. NSC power and influence would continue to expand throughout the era covered by this study. By 1972, Jack Burke, editor for the organization's periodical, *Farm Safety Review*, indicated that "the organization has grown from small potatoes to one of 400-plus employees and a $11 million budget"; see Jack Burke, "NSC's Sixtieth Year," *Farm Safety Review*, November/December 1972, 2.

11. Dennis Murphy, "President's Message to the Summer Meeting of the National Institute for Farm Safety, Inc." (speech given at the Summer Meeting of the National Institute for Farm Safety, Orlando, Florida, June 16, 1986), 7, Special Collections, Iowa State University Library, Iowa State University, Ames.

12. *Transactions: National Safety Congress* (Chicago: National Safety Council, 1938), 33; *Transactions: National Safety Congress* (Chicago: National Safety Council, 1940), 31; C. L. Hamilton, "Agriculture's Safety Challenge," *Agricultural Engineering* 46 (April 1945): 145.

13. "New Farm Safety Leaders," *Farm Safety Review*, January/February 1944, 2; "Staff Changes," *Farm Safety Review*, July/August 1954, 15; "NSC Appoints L. H. Hodges Vice President for Farms," *Farm Safety Review*, November/December 1971, 3; "Pfister Heads NSC Farm Conference," *Farm Safety Review*, November/December 1974, 3.

14. G. L. Noble, "Agriculture's Safety Movement," *Farm Safety Review*, November/December 1946, 15; "We Couldn't Do It Without Investors," *Farm Safety Review*, March/April 1960, 14–15. For similar listings of benefactors, see "Investors in Farm Safety," *Farm Safety Review*, May/June 1952, 15; "Investors in Farm Safety," *Farm Safety Review*, July/August 1953, 15; "Investors in Farm Safety," *Farm Safety Review*, May 1955, 15; Howard Pyle "NSC Salutes You," *Farm Safety Review*, September 1963, 3.

15. "M.S.A. Sponsors Review," *Farm Safety Review*, January/February 1951, 13; "M.S.A. Renews Review Sponsorship," *Farm Safety Review*, March/April 1952, 3; "A.S.A.E. Gets Review," *Farm Safety Review*, July/August 1953, 4. The *Review* attracted wide readership from those in the agricultural business community. For instance, as early as 1947, "fourteen thousand implement dealers" and "nine thousand extension leaders" received the periodical; see "How You Get This Magazine," *Farm Safety Review*, March/April 1953, 4.

16. Maynard Coe, "The Reasons for National Farm Safety Week," *Farm Safety Review*, July/August 1944, 3. The USDA also had stressed the importance of the farm accident issue by creating its own safety council consisting of representatives from the agency's many branches. The USDA Safety Council met once each month and augmented the NSC publication of safety materials. It also aided the National Safety Week effort by distributing safety information to state extension offices. For a brief discussion of this committee and its activities, see *Farm Safety Review*, November/December 1951, 13.

17. "Lehman Reports on National Farm Safety Week," *Farm Safety Review*, September/October 1944, 4. Lehman's occupational status can be found in "Agricultural Safety: Panel Discussion," *Transactions: National Safety Congress* (Chicago: National Safety Council, 1940), 43.

18. "Agricultural Leaders Endorse Farm Safety Week, July 21–27," *Farm Safety Review*, May/June 1947, 5; "Agricultural Leaders Endorse Farm Safety Week's Objective," *Farm Safety Review*, July/August 1948, 3.

19. "National Farm Safety Week on the Air," *Farm Safety Review*, July/August 1951, 3; Maynard H. Coe "Farm Division Meeting," *Farm Safety Review*, January/February 1955, 12.

20. "The Week for Rural America," *Farm Safety Review*, May 1962, 4.

21. Jack Burke, "Farm Safety Week, 1965," *Farm Safety Review*, May/June 1965, 2.

22. "Farm Safety at the Congress," *Farm Safety Review*, December 1955, 4.

23. "Committee Plans Youth Sessions," *Farm Safety Review*, March 1957, 3.

24. "Youth Looks to the 60s," *Farm Safety Review*, November/December 1959, 3.

25. "Farm Safety at the Congress," *Farm Safety Review*, December 1955, 4.

26. "For 1956 Safety Congress," *Farm Safety Review*, March 1956, 15; "Committee Plans Youth Sessions," *Farm Safety Review*, March 1957, 3.

27. "Farm Safety at the Congress," 5; "For 1956 Safety Congress," 15.

28. Hurt, *American Agriculture*, rev. ed., 189–192. Hurt describes how the USDA's size and scope continued to expand throughout the twentieth century, becoming by the 1990s the fourth largest government spender and the sixth largest government employer, with more assets than all but three U.S. corporations; see Hurt, *American Agriculture*, rev. ed., 393.

29. In 1948, the *Farm Safety Review* reported that officials for the USDA's Department of Economics were engaged in a farm accident survey; see "USDA Survey," *Farm Safety Review*, July/August 1948, 13.

30. "USDA Safety Council," *Farm Safety Review*, November/December 1951, 13.

31. Norval Wardle (presentation to Iowa Farm Safety Council Board of Directors, November 11, 1955), Iowa Farm Safety Council Records, Special Collections, Iowa State University Library, Iowa State University, Ames.

32. Lloyd H. Davis, "USDA's Safety Program," *Farm Safety Review*, November/December 1971, 5–6. Davis specified that the task force was publishing material informing the farm public regarding OSHA standards, intensifying cooperative work with the NSC and other organizations, and improving farm accident data collection methods; for additional details regarding both the structure and functions of the task force, see Clifford Hardin, *Secretary's Memorandum No. 1723: Safety in American Agriculture* (Washington, D.C.: U.S. Department of Agriculture, Office of the Secretary, 1971).

33. Plambeck, ed., *Iowa Farm Safety in the Twentieth Century*, 156, 140.

34. Cochrane, *Development of American Agriculture*, 105–107; Jim Hightower, *Hard Tomatoes, Hard Times: A Report of the Agribusiness Accountability Project on the Failure of America's Land-Grant College Complex* (Cambridge, Massachusetts: Schenkman Publishing Company, 1973), 152–155.

35. Early pioneers of Iowa Extension, such as Ralph Bliss, and farm safety specialists, such Norval Wardle, were both educated at Iowa State College. Land-grant colleges were instrumental in training such specialists in other states as well. For a discussion of Ralph Bliss's background and education, see Dorothy Schwieder, "The Iowa State College Cooperative Extension Service through Two World Wars," *Agricultural History* 64, no. 2 (1990): 225.

36. Cochrane, *The Development of American Agriculture*, 6.

37. *Born in the Country*, 172–173.

38. Schwieder, "The Iowa State College Cooperative Extension Service through Two World Wars," 220.

39. Danbom, *Born in the Country*, 174.

40. Ibid., 221.

41. Ibid.

42. Ibid., 219. For examples of the many ways extension employees assisted in the nation's mobilization efforts, particularly regarding wartime agricultural labor shortages in states such as Maryland, California, and Kentucky, see T. B. Symons, "To Meet Labor Shortage," *Extension Service Review*, May 1942, 77; "Training New Hired Men," *Extension Service Review*, May 1943, 69; and B. H. Crocheron, "How Many Workers," *Extension Service Review*, July 1944, 100. Extension's influence increased in the following decade. In 1954, Ezra Taft, U.S. secretary of agriculture, indicated that "extension now reaches about 85 percent of all farm families

in the nation"; see Ezra Taft Benson, "The Role of Extension," *Extension Service Review*, January 1954, 3.

43. David Danbom, while commenting on the emergence of Cooperative Extension, stated that "the popularity of extension in the South excited those who saw it as a potential means for reforming agriculture everywhere in the country." See Danbom, *Born in the Country*, 174.

44. Fitzgerald, *Every Farm a Factory*, 21, 24–27, 43–44.

45. Danbom, *Born in the Country*, 172, 173, 176; Jellison, *Entitled to Power*, 16.

46. Danbom, *Born in the Country*, 174.

47. Jellison, *Entitled to Power*, 36.

48. M. L. Wilson, "A New Chapter Begins," *Extension Service Review*, January 1946, 1.

49. Martha Kohl, "County in Virginia Wins Highway Safety Contest," *Farm Safety Review*, December 1956, 5; "Hoosier Women Win Lane Award," *Farm Safety Review*, March/April 1959, 7; *Transactions: National Safety Congress* (Chicago: National Safety Council, 1951), 25; "Women at Work," *Farm Safety Review*, July/August 1959, 13.

50. Hurt, *American Agriculture*, rev. ed., 258–259.

51. Thomas Wessel and Marilyn Wessel, *4-H: An American Idea, 1900–1980* (Chevy Chase, Maryland: National 4-H Council, 1982), 36, 37, 76, 88.

52. Gertrude L. Warren, "A Brief Review: The 4-H Story," *Extension Service Review*, May 1955, 93.

53. Ralph E. Bender, Robert E. Taylor, Chester K. Hansen, and L. H. Newcomb, *The FFA and You* (Danville, Illinois: Interstate Printers and Publishers, Inc., 1979), 7; *The Iowa FFA: For Fifty Years* (Des Moines: Iowa Association FFA, Career Education Division, Department of Public Instruction, 1978), vi, 1.

54. *The Iowa FFA: For Fifty Years*, vi.

55. Bender, Taylor, Hansen, and Newcomb, *The FFA and You*, 620.

56. *The Iowa FFA: For Fifty Years*, vi, 2.

57. John Morton Blum, *V Was for Victory* (New York: Harcourt Brace Jovanovich, 1976), 102, 197, 144. For a brief discussion of wartime mobilization, see Paul S. Boyer, *Promises to Keep: The United States since World War II* (New York: Houghton Mifflin Company, 2005), 7–16.

58. Judy Barrett Litoff and David C. Smith, "To the Rescue of the Crops: The Women's Land Army during World War II," *Prologue* 25 (Winter 1993): 347.

59. Schwieder, "The Iowa State College Cooperative Extension Service through Two World Wars," 220.

60. Litoff and Smith, "To the Rescue of the Crops," 348. In the early summer of 1943, Francis Flood, a high-ranking War Food Administration official, indicated that obtaining enough workers to supply the nation's agricultural sector was "one of the biggest single assignments extension has ever had in its history of big

assignments"; see Francis Flood, "The Challenge of the 99 Percent," *Extension Service Review*, June 1943, 81.

61. "A Letter from the President of the United States: To the 4-H Club Members of the United States," *National 4-H Club News*, April 1942, 5.

62. Alan Clive, "The Michigan Farmer in World War II," *Michigan History* 60 (Winter 1976): 296.

63. "Rural Youth Are Aid to Defense in Five Ways," *Better Iowa*, December 15, 1941, 1; "Mobilization Week Feb. 6–14 Wanted–Two Million or More Rural Boys and Girls to Enroll in the 4-H Victory Program," *National 4-H Club News*, January 1943, 4.

64. H. B. Walker, "Agricultural Engineers on the Food Front," *Agricultural Engineering* 24 (January 1943): 5, 7.

65. Litoff and Smith, "To the Rescue of the Crops," 349–355.

66. Wesley Longnecker, interview by Derek Oden, digital recording, January 2, 2013.

67. Litoff and Smith, "To the Rescue of the Crops," 349, 356; "70,000 Workers Placed by Farm Labor Offices," *Better Iowa*, October 1, 1945, 1. For more on the activities of Iowa's Extension Service to ensure adequate labor supplies, see "Farmers Can't Lose More Labor," *Wallaces Farmer and Iowa Homestead*, July 3, 1943, 4.

68. Maynard Coe, "Boom in Farm Safety," *Farm Safety Review*, March/April, 1944, 2. See also Arnold P. Yerkes, "Engineering in Wartime Agriculture," *Agricultural Engineering* 23 (April 1942): 117–125, as an example of the effort to improve the overall efficiency of the agricultural sector as part of the nation's plans for victory.

69. "Maynard Coe to New Post," *Farm Safety Review*, March/April, 1964, 13.

70. E. W. Lehmann and H. P. Bateman, "Improving Agricultural Efficiency of Farm Labor to Meet Wartime Demands," *Agricultural Engineering* 23 (September, 1942): 280; H. L. Miner "Food Is a Weapon of War," *Farm Safety Review*, January/February 1943, 3.

71. *Better Iowa*, March 23, 1942, 1.

72. George M. Gehant, "A County Agent Discusses Farm Safety," *Farm Safety Review*, September/October 1943, 2. In 1943, WHO radio located in Des Moines, Iowa, featured an announcement they had received from the NSC's farm division staff. The message summarized a key reason why the farm safety movement had emerged: The narrator proclaimed that safety-conscious farmers were "saving farm manpower for war power"; see Elmer Lotstrom and Jim Chapman, "Farm Safety on the Air Waves," *Farm Safety Review*, March/April, 1943, 8.

73. "'D' Day on the Farm Front," *Farm Safety Review*, July/August, 1944, 15.

74. Claude R. Wickard, "Safety in Wartime Farming," *Farm Safety Review*, July/August, 1943, 4–5; Florence L. Hall, "Safety Tips for Women's Land Army," *Farm*

Safety Review, May/June 1945, 12; "Improve Farm Drinking Facilities," *Farm Safety Review*, September/October 1943, 13.

75. Plambeck, ed., *Iowa Farm Safety in the Twentieth Century*, 26–28. For a brief but useful discussion of the activities of farm safety councils and committees in other states, see "Illinois Committee Meets," *Farm Safety Review*, January/February 1944, 13.

76. Maynard Coe, "A Decade of Farm Safety," *Farm Safety Review*, March/April 1954, 12–15; "State Organization Farm Safety," *Farm Safety Review*, January/February 1947, 6.

77. "Farm Safety Roundup," *Farm Safety Review*, July/August 1947, 13; "The Farm Safety Story," *Farm Safety Review*, May/June 1960, 8.

78. "Farm Safety Roundup," 12–15; "The Farm Safety Story," 7.

79. Plambeck, ed., *Iowa Farm Safety in the Twentieth Century*, 9.

80. "Indiana Officers Change," *Farm Safety Review*, March/April 1948, 13.

81. Richard A Tjalma, "Institute of Agricultural Medicine in Iowa," *Public Health Report, National Institutes of Health* 73, no. 7 (July 1958): 616.

82. Ibid., 616–618.

83. L. W. Knapp to Norval Wardle, March 4, 1963, Norval Wardle Collection, Special Collections, Iowa State University, Ames. Such collaboration between the faculty of Iowa State University and the staff of Institute of Agricultural Medicine was sometimes more formal and on a much larger scale. For instance, in 1966 the *Ames Daily Tribune* reported that both organizations were involved in a pesticide study examining the health impacts of agricultural chemicals. The study was financially supported by the U.S Public Health Service. See "ISU, SUI Team Up on Pesticide Study," *Ames Daily Tribune* (Iowa), June 9, 1966, 1.

84. Bruce A. Harrison, "University's Farm Accident Building Nears Completion," *Cedar Rapids Gazette* (Iowa), August 7, 1966, 19b.

85. "UI Accident Prevention Lab to Be Dedicated," *Iowa City Press-Citizen* (Iowa), October 1, 1968, 3A.

86. Harrison, "University's Farm Accident Building Nears Completion," 19b.

87. Phyllis Fleming, "Isolated Oakdale to Become Hub of Science," *Cedar Rapids Gazette* (Iowa), November 7, 1965, 1b.

88. Harrison, "University's Farm Accident Building Nears Completion," 19b.

89. "UI Accident Prevention Lab to Be Dedicated," 3A.

90. L. W. Knapp, "We Need More Research: Rural Accident Prevention Lags Behind Technological Advances in Other Areas," *Farm Safety Review*, March/April 1964, 11.

91. Harrison, "University's Farm Accident Building Nears Completion," 19b.

92. Beth Linker, *War's Waste: Rehabilitation in World War I America* (Chicago: University of Chicago Press, 2014), 2.

93. "Barry Named to Pollution Control Unit," *Iowa City Press-Citizen*, September 9, 1968, 5a; "Three to Attend Rural Medicine Parley in Tokyo," *Iowa City*

Press-Citizen, September 27, 1969, 2a. For examples of Clyde Berry's articles in *American Journal of Public Health*, see "Agricultural Industry: The Changing Pattern," *American Journal of Public Health and the Nation's Health* 49, no. 5 (May 1959): 616–621; "Organized Research in Agricultural Health and Safety," *American Journal of Public Health and the Nation's Health* 56, no. 7 (May 1966): 424–428; and "Rural Employment," *American Journal of Public Health* 61, no. 12 (December 1971): 2474–2476.

94. National Safety Council, Farm Department, "Top Spots to Hodges and Knapp," *Agri-Safety News*, no. 4 (1971): 1, Norval Wardle Collection, Special Collections, Iowa State University, Ames.

95. L. W. Knapp, "Farm Conference Chairman," *Farm Safety Review*, November/December 1971, 4.

96. "Benton County's Success Story," *Farm Safety Review*, November/December 1959, 8. See "First Radio Farm Accident Survey," *Farm Safety Review*, September/October 1950, 4–5; *Transactions: National Safety Congress* (Chicago: National Safety Council, 1952), 14–15; Philip Tichener, "Olmsted County, Minn., Organizes a Hazard Hunt," *Extension Service Review*, November 1956, 207.

97. Don Muhm, "U.S. Could Act if Farm Tractors Not Made Safer," *Des Moines Register* (Iowa), December 6, 1969, 1.

98. "Iowa Safety Council Meeting Minutes," Des Moines, Iowa, Iowa Safety Council, January 19, 1973, Iowa Safety Council Collection, Special Collections, Iowa State Library, Ames.

CHAPTER 6

1. Maynard Coe, "Call a Specialist," *Farm Safety Review*, November/December 1944, 1. In 1950, a *Farm Safety Review* writer indicated that in the state of Wisconsin "the farm safety specialist and his staff work as stimulators and coordinators for farm safety activities throughout the state." See "Falling Fatalities: Wisconsin Program Cuts Accidental Deaths," *Farm Safety Review*, March/April 1950, 3. Prior to Maynard Coe's appointment as the head of the National Safety Council's farm division in 1944, he had served as the 4-H Club Leader at Kansas State University; see "Maynard Coe to New Post," *Farm Safety Review*, March/April, 1964, 13.

2. Maynard Coe, "Specialists Needed," *Farm Safety Review*, November/December 1952, 1.

3. A. J. Schwantes, "Farm Safety Opportunities of Agricultural Colleges," *Transactions: National Safety Council* (Chicago: National Safety Council, 1951), 5.

4. Thomas P. Hughes, *American Genesis: A Century of Invention and Technological Enthusiasm, 1870–1970* (New York: Viking, 1989), 4.

5. "Farm Safety Roundup," *Farm Safety Review*, November/December 1949, 14;

"Illinois Gets Specialist," *Farm Safety Review*, September/October 1946, 6; "Farm Safety Roundup," *Farm Safety Review*, July/August 1949, 14.

6. Danbom, *Born in the Country*, 167–168.

7. Fitzgerald, *Every Farm a Factory*, 6.

8. Ibid., 21.

9. Ibid., 6, 18, 21.

10. Harold E. Pinches, "The Agricultural Engineer Grows Up," *Agricultural Engineering* 18 (February 1937): 47.

11. "Agricultural Safety: Panel Discussion," *Transactions: National Safety Congress* (Chicago: National Safety Council, 1940), 43.

12. C. L. Hamilton, "Engineering Farm Safety," *Agricultural Engineering* 27 (June 1946): 260.

13. "Responsibilities of Colleges in Farm and Home Safety," *Transactions: National Safety Council* (Chicago: National Safety Council, 1953), 23.

14. A. J. Schwantes, "Farm Safety Education in Colleges," *Transactions: National Safety Congress* (Chicago: National Safety Council, 1954), 18.

15. U.S. Department of Agriculture, Bureau of Agricultural Economics, *The Prevention of Accidents and in Homes* (Washington, D.C: U. S. Department of Agriculture, 1942), 21.

16. Frank J. Zink, "Farm Equipment Can Be Operated Safely" (address delivered at the Nebraska Conference on Home and Farm Safety, Lincoln, October 2, 1942), 3, Special Collections, Iowa State University, Ames.

17. Ibid., 6.

18. Ibid., 7.

19. "Report to the President," *Farm Safety Review*, July 1956, 13; "Farm Safety Clinic: Discussion Groups Air Ideas to Reduce Accidents," *Farm Safety Review*, August 1956, 4–6.

20. "Farm Safety Round Up," *Farm Safety Review*, July/August 1945, 15.

21. "Five State Specialists in Farm Safety Meet," *Farm Safety Review*, July/August 1946, 11.

22. B. G. Fowler, "Oregon Mobilizes for Safety," *Farm Safety Review*, July/August 1948, 5.

23. Schwantes, "Farm Safety Opportunities of Agricultural Colleges," 6.

24. "The Farm Safety Story," *Farm Safety Review*, March 1956, 3.

25. Harry M. Pontius, "A Career in Safety?" *Farm Safety Review*, March/April 1959, 3.

26. "The Farm Safety Story," *Farm Safety Review*, May/June 1960, 7.

27. Despite their small number, farm safety experts had established the National Institute of Farm Safety as an autonomous nonprofit organization by 1962. The story of their separation from the NSC will be recounted in chapter 9.

28. Mark A. Purschwitz and William E. Field, "Safety Spending: Is Agriculture

Being Shortchanged?" *Agricultural Engineering* 69 (November/December 1988): 10.

29. Ibid., 10, 18.

30. Fenton, "Agricultural Engineering in a Period of Confusion," 10, 18.

31. Ophthalmologists became concerned about the escalating number of anhydrous ammonia accidents during the late 1960s, which then often caused blindness; Plambeck, ed., *Iowa Farm Safety in the Twentieth Century*, 116.

32. Here are a few examples. In the late 1960s, William B. Buck, an Iowa State University professor of veterinary medicine, documented the destructive consequences of the careless use of chemical pest treatments for livestock. In the late 1970s, ISU entomology professor Dr. David Mick authored numerous publications on methods of preventing agricultural chemical accidents and taught farm chemical safety classes; see William B. Buck, unpublished report regarding livestock poisoning episodes in Iowa during 1969, 1–2, Special Collections, Iowa State University, Ames. See also Plambeck, ed., *Iowa Farm Safety in the Twentieth Century*, 71.

33. Plambeck, ed., *Iowa Farm Safety in the Twentieth Century*, 21.

34. Leon J. Urban, National Safety Council, Manager, Farm Department, to Norval Wardle, July 18, 1973, Norval Wardle Collection, Iowa State University Library, Iowa State University, Ames. The meeting was cooperatively sponsored by the National Safety Council, the U.S. Department of Agriculture (USDA), the American Farm Bureau Federation, and the Farm and Industrial Equipment Institute.

35. James D. Hill, "Careers in Safety," *Transactions: National Safety Council* (Chicago: National Safety Council, 1957), 27.

36. "Stuckey Farm Conference Chairman," *Farm Safety Review*, November/December 1968, 3.

37. Ralph E. Patterson, "We Are Responsible for Farm Safety," *Extension Service Review*, November 1965, 14.

38. National Institute of Farm Safety, Professional Improvement Committee Report, June 18, 1970, Norval Wardle Collection, Special Collections, Iowa State University Library, Iowa State University, Ames, Iowa. This report and similar documents reveal how NIFS leaders created a number of committees, including those for Tractor and Machinery, Fire and Electricity, Emergency Preparedness, Home and Farmstead, Rural Traffic, Farm Chemicals, and others associated with the study and reduction of agricultural dangers. Committees dedicated to improving the organization's effectiveness and prominence included the membership, program planning, and professional development committees. For details regarding who served on these groups, see National Institute of Farm Safety, Program Area Committee Roster 1967/1968, Norval Wardle Collection, Special Collections, Iowa State University Library, Iowa State University, Ames.

39. Robert A. Aherin, "Professional Development and You" (paper presented at

the fall National Institute of Farm Safety Meeting, location not provided, October 20, 1981), 2.

40. Ibid., 3.

41. Robert A. Aherin to NIFS members, August 8, 1980, Norval Wardle Collection, Special Collections, Iowa State University Library, Iowa State University, Ames.

42. Robert A. Aherin to NIFS Professional Improvement Committee members, September 25, 1981, Norval Wardle Collection, Special Collections, Iowa State University Library, Iowa State University, Ames.

43. "New State Leaders," *Farm Safety Review*, March/April 1946, 16; "Farm Safety Roundup," *Farm Safety Review*, November/December 1949, 14; "Farm Safety Clinics," *Farm Safety Review*, May/June 1950, 16. Farm safety specialists also found employment with state Farm Bureau organizations, agricultural associations, and a variety of other groups. For example, see "Illinois Gets Specialist," *Farm Safety Review*, September/October 1946, 6; and "Ohio Employs Farm Safety Specialist," *Farm Safety Review*, November/December 1944, 15.

44. "Back to Farm Safety," *Farm Safety Review*, March 1956, 11; Plambeck, ed., *Iowa Farm Safety in the Twentieth Century*, 150; Donald Jensen, "National Institute for Farm Safety Report, August 1975," Norval Wardle Collection, Special Collections, Iowa State University Library, Iowa State University, Ames; "Former County Agent Announces Retirement," *The Post Crescent* (Appleton, Wisconsin), April 29, 1967, 9; "Hogsett Obituary" tributes.com, www.tributes.com/show/Ordie-L.-Hogsett-95758767; "New State Leaders," *Farm Safety Review*, March/April 1946, 16.

45. "Whitlock Gets Specialist," *Farm Safety Review*, September/October, 1946, 7.

46. B. G. Fowler, "Oregon Mobilizes for Safety," *Farm Safety Review*, July/August 1948, 5.

47. "Pennsylvania Gets Safety Specialist," *Farm Safety Review*, January/ February 1948, 13.

48. "Farm Safety Roundup: New State Farm Safety Directors," *Farm Safety Review*, November/December 1950, 14.

49. For examples illustrating these diverse tasks, relating to both livestock and machines, see Swanson, "The Dairy Bull—Dr. Jekyll and Mr. Hyde," 598. For examples of efforts to reduce accidents related to all manner of agricultural equipment, including tractors, harvesting equipment, and crop transport devices, see "The Man Behind the Wheel," *Farm Safety Review*, May/June 1961, 7; Norval J. Wardle, "Half Blind or Full Vision," *Farm Safety Review*, March/April 1969, 12–13.

50. Farm safety specialists worked to meet the challenge of the persistent lack of farm accident data by contributing to the collection and analysis of farm accident statistics; for articles discussing the persistent need for data regarding chemical accidents, see William J. Fletcher, "Pesticide Worker Accidents" (paper

presented at the National Conference on Protective Clothing and Safety Equipment for Pesticide Workers, Centers for Disease Control and Prevention, Atlanta, Georgia, May 1–2, 1972), 1, 4; and National Safety Council, *Rural Accident Prevention Bulletin: Safe Use of Pesticides* (Chicago: National Safety Council, 1976), 1.

51. Robert D. Schnieder and Robert J. Florell, "Farm Tractor Upsets Are Man Killers," *Farm Safety Review*, May/June 1969, 3; "AMA Calls for Tractor Protection," *Farm Safety Review*, September/October 1967, 15.

52. Plambeck, ed., *Iowa Farm Safety in the Twentieth Century*, 25.

53. Norval Wardle to Randall C. Swanson, November 26, 1969, Norval Wardle Collection, Special Collections, Iowa State University Library, Ames. See also "Gets Specialist, Iowa Joins Ranks," *Farm Safety Review*, September/October 1947, 10.

54. Norval Wardle, professional résumé, 1964, Norval Wardle Collection, Special Collections, Iowa State University Library, Iowa State University, Ames.

55. Plambeck, ed., *Iowa Farm Safety in the Twentieth Century*, 149. Longtime Illinois farm safety specialist O. L. Hogsett, like Wardle, also had a personal reason for his farm safety calling. In 1943, he suffered a corn-picker accident that resulted in severe injury to both of his hands; see "12th National Farm Safety Week," *Farm Safety Review*, July 1955, 3.

56. Norval Wardle, professional résumé, 1964, 1–2.

57. Norval Wardle to James L. Williams, February 27, 1974, Norval Wardle Collection, Special Collections, Iowa State University Library, Iowa State University, Ames.

58. Plambeck, ed., *Iowa Farm Safety in the Twentieth Century*, 16, 20–21.

59. National Safety Council, Farm Department, *Agri-Safety News*, 1971, 2, Norval Wardle Collection, Special Collections, Iowa State University Library, Iowa State University, Ames.

60. "Twelve Days: With Stuckey, It Was Always Safety First," *From Woody's Coach: Our Playbook on OSU History*, Ohio State University Libraries, December 11, 2013, https://library.osu.edu/blogs/archives/2013/12/11/twelve-days-with-stuckey-it-was-always-safety-first/; and "Stuckey Farm Conference Chairman," *Farm Safety Review*, November/December 1968, 3.

61. "Appointments in Nebraska and Missouri," *Farm Safety Review*, March/April 1954, 10.

62. William Splinter, ed. "The Biological Systems Engineering Program in Nebraska 1895–2011," Biological Systems Engineering, Digital Common University of Nebraska–Lincoln, January 1, 2012, http://digitalcommons.unl.edu/cgi/viewcontent.cgi?article=1212&context=biosysengfacpub.

63. "New Safety Specialist," *Farm Safety Review*, November/December 1949, 14.

64. "New State Leaders," *Farm Safety Review*, March/April 1946, 16.

65. W. T. Spanton, "Farm Safety in Agricultural Education," *Farm Safety Review*, November/December, 1944, 10–11.

66. Advocates of Vo-Ag education had been especially active and successful in establishing such instruction in the nation's secondary schools during the Progressive Era. For an old but useful survey of the origins of agricultural education in the United States, see Alfred Charles True, *A History of Agricultural Education in the United States, 1785–1925* (New York: Arno Press, 1929), 436.

67. Maynard Coe, "Specialists Needed," *Farm Safety Review*, November/December 1951, 1.

68. Glen Hetzel, "National Institute of Farm Safety President's Address" (speech given at NIFS meeting, St. Louis, Missouri, Summer 1985), Norval Wardle Collection, Iowa State University Archives, Iowa State University, Ames.

69. "Hogsett Named Illinois Specialist," *Farm Safety Review*, March/April 1955, 12.

70. "Michigan Gets Farm Safety Specialist," *Farm Safety Review*, May/June 1946, 10.

71. "Stevens Fills Missouri Post," *Farm Safety Review*, September/October 1960, 14.

72. Plambeck, ed., *Iowa Farm Safety in the Twentieth Century*, 22.

73. Randall C. Swanson, "Wisconsin Outlaws Accidents," *Farm Safety Review*, September/October 1944, 5.

74. "Illinois Gets Specialist," *Farm Safety Review*, September/October 1946, 7.

75. "New Appointments," *Farm Safety Review*, November/December 1953, 3.

76. "Safety Specialist for Michigan," *Farm Safety Review*, September/August 1949, 14; "Farm Safety Round-up: Steinicke Resigns Michigan Post," *Farm Safety Review*, August 1955, 14; "Pfister Appointed Safety Michigan Specialist in Michigan," *Farm Safety Review*, August 1955, 14.

77. "The Formative Years of NIFS," *International Society for Agricultural Safety and Health*, http://isash.org/resources-2/historical-papers/early-development-of-farm-safety/.

78. "Safety Experts Meet," *Farm Safety Review*, July/August 1946, 3.

79. "Farm Safety Institute," *Farm Safety Review*, August 1955, 5.

80. Grice E. Sexton to Norval Wardle, June 25, 1956, Norval Wardle Papers, Special Collections, Iowa State University Library, Ames, Iowa.

81. "Recent Programming Aids," *Farm Safety Review*, March/April 1961, 5.

82. "4-H for Safety," *Farm Safety Review*, January/February 1947, 3; "For Your Operator Training Program," *Farm Safety Review*, July/August 1961, 12; Jack Burke, "Beaver Dam's FFA: A Safety Program for Young and Old," *Farm Safety Review*, September/October 1967, 3.

83. Norval Wardle, "How the Safety Specialist Multiplies His Efforts," *Extension Service Review*, March 1963, 51.

84. For examples of rural youth safety programs, see William D. Lutes, "4-H Tractor Clubs Build Safe Drivers," *Farm Safety Review*, November/December 1954, 8; "Rural Schools Check Farm Fires," *Farm Safety Review*, July/August 1952, 11; "Let the Sun Shine," *Farm Safety Review*, November/December 1969, 13.

85. Chuck Worcester, "Three Strikes: A Farm Safety Home Run," *Farm Safety Review*, November/December 1947, 4–5.

86. W. E. Stuckey, "4-H Safety Ways: Safety Is a Family Affair," *Farm Safety Review*, January 1957, 21.

87. Wilbur E. Stuckey, "Position Description of Wilbur E. Stuckey," Norval Wardle Collection, Iowa State University Library, Iowa State University, Ames.

88. Ibid.

89. "Farm Safety—A Sound 'Investment,'" *Farm Safety Review*, November/December 1964, 4.

90. Iowa Farm Safety Council, "Annual Minutes, March 12, 1954" (Ames, Iowa, Iowa Farm Safety Council Records), 1–2, Special Collections, Parks Library, Iowa State University, Ames.

91. Kenneth K. Barnes, "Key Items You May Have Missed in the DOT Report," *Implement and Tractor*, July 1971, 35.

CHAPTER 7

1. *Transactions: National Safety Council* (Chicago: National Safety Council, 1961), 76.

2. Jack Burke, "Out of That Mountain, You Could Mine Some Gold," *Agri-Safety Newsletter*, Fall 1978, 2.

3. "Electronic Safety Quiz," *Farm Safety Review*, September/October 1949, 10.

4. J. S. Jones, "Why Farm Organizations Should Be Interested in Farm Safety," *Farm Safety Review*, November/December 1944, 6.

5. Plambeck, ed., *Iowa Farm Safety in the Twentieth Century*, 9; Edward S. Adams to Mr. Harry M. Pontius, February 25, 1954, Iowa Safety Council Collection, Special Collections, Iowa State Library, Ames.

6. J. D. Armstrong to Iowa Farm Mutual Insurance Company, July 9, 1953, Iowa Safety Council Collection, Special Collections, Iowa State Library, Ames; Edward S. Adams to Harry L. Gross, April 8 1955, 1, Iowa Safety Council Collection, Special Collections, Iowa State Library, Ames.

7. Maynard Coe, "Indifference," *Farm Safety Review*, January/February 1951, 1.

8. *Transactions: National Safety Congress* (Chicago: National Safety Council, 1951), 32.

9. C. W. Dalbey, "Human Resources Our Most Important Commodity," *Farm Safety Review*, September/October 1968, 11.

10. Don Rimathe, interview by Derek Oden, digital recording, January 3, 2013.

11. Pat Larkin, interview by Derek Oden, digital recording, December 29, 2012.

12. Jim Klein, interview by Derek Oden, digital recording, December 15, 2012.

13. For an example regarding how authors of safety textbooks treated the farm safety problem, see Harry W. Gentles and George H. Betts, *Habits for Safety: A Textbook for Schools* (New York: Bobbs-Merrill Company, 1937), 110–124.

14. "New Safety Info," *Farm Safety Review*, September/October 1973, 11.

15. "Farm Safety Week," *Farm Safety Review*, May/June 1974, 7.

16. By the late 1970s, the NSC was publishing the *Agri-Safety Newsletter* in addition to the *Farm Safety Review*. This publication was less substantial than the *Review* and aimed more exclusively toward safety professionals. The newsletter frequently advertised safety resources that addressed current safety topics.

17. Charles R. Cavagnaro, "A Full Measure of Safety for Young Farmhands," *Farm Safety Review*, March/April 1965, 10.

18. U.S. Department of Agriculture in cooperation with National Chemicals Association, National Safety Council, U.S. Department of Health Education and Welfare, National Vocational Agricultural Teachers Association, Manufacturing Chemists' Association, Inc., *Chemicals in Agriculture: A Guide for Adult and Youth Programs in Agricultural Chemical Safety* (Washington, D.C., 1964), 3; "Four Keys to Pesticide Safety," *Farm Safety Review*, November/December 1965, 6.

19. "Farm Safety Roundup," *Farm Safety Review*, September/October 1943, 14.

20. Stanley H. Gaines, "Prevention Is the Key to Farm Safety: The USDA and Farm Safety," *Farm Safety Review*, July/August 1947, 6; U.S. Department of Agriculture et al., *Chemicals in Agriculture*, 3.

21. U.S. Department of Agriculture, Bureau of Agricultural Economics, *The Prevention of Accidents on Farms and in Homes* (Washington, D.C., 1942).

22. For an additional example, see Harvey Hirning, Robert Woell, and David Swenson, *Electric Farm Power: Look Up!* (Fargo: North Dakota State Agricultural Engineering Department, Cooperative Extension Service, and the North Dakota Power Council, 1977), 1–4.

23. *Transactions: National Safety Congress* (Chicago: National Safety Council, 1964), 26.

24. "NSC Award Goes to Magazine and Radio Stations," *Farm Safety Review*, January/February 1950, 3.

25. "For Exceptional Service to Farm Safety Program," *Farm Safety Review*, January/February 1953, 8.

26. Milon Grinnell, "Present Positive Programs," *Farm Safety Review*, January/February 1959, 3.

27. *Transactions: National Safety Council* (Chicago: National Safety Council, 1953), 41–42.

28. U.S. Department of Agriculture et al., *Chemicals in Agriculture*, 3–4.

29. "4 Keys to Pesticide Safety," *Farm Safety Review*, November/December 1965, 6.

30. "Timely Visual Aids," *Farm Safety Review*, September/October 1952, 3.

31. Ibid. See also "The Safest Township Anywhere," *Farm Safety Review*, January/February 1961, 5, which discusses the content of a film sponsored by General Motors.

32. Rex Peterson, interview by Derek Oden, digital recording, January 1, 2013.

33. "Farm Radio Audience in Pennsylvania," *Farm Safety Review*, March/April 1943, 6.

34. Dan Thompson, "What You Can Do to Help Put Safety on the Air," *Farm Safety Review*, July/August 1943, 6.

35. "Farm Safety on the Air," *Farm Safety Review*, March/April 1944, 6.

36. "Farm Week Publicity Winners Named," *Farm Safety Review*, November/December 1946, 13; "NSC Award Goes to Magazine and Radio Stations," 3.

37. "1951 Publicity Awards," *Farm Safety Review*, January/February 1952, 9; see also "Safety Publicity Winners," *Farm Safety Review*, January/February 1949, 8, for an earlier example of an awards banquet sponsored by NSC.

38. "Focus on People," *Farm Safety Review*, November/December 1967, 3.

39. "Safety Rhythm," *Farm Safety Review*, May/June 1948, 3; "Use Radio to Promote Farm Safety," *Farm Safety Review*, November/December 1948, 12; Chuck Worcester, "Gadgets That Save Lives," *Farm Safety Review*, March/April 1949, 3.

40. Bernard Stodghill, interview by Derek Oden, digital recording, December 29, 2012.

41. Chester S. Hutchison, "Jingles and Slogans Promote Safety," *Farm Safety Review*, November/December 1949, 11.

42. Frank F. Altwood, "Radio Contest Saves Lives," *Farm Safety Review*, September/October 1950, 13.

43. Wallace N. Dudney, "Publicity Techniques for Farm Safety Leaders," *Farm Safety Review*, June 1956, 12–13.

44. Miriam J. Kelley, "Open the Door to Television," *Extension Service Review*, July 1953, 129.

45. "Farm Safety Roundup," *Farm Safety Review*, July/August 1945, 14.

46. "Farm Safety via TV in Oklahoma," *Farm Safety Review*, July/August 1952, 4.

47. Ibid. "Selling Safety with Television in Illinois," *Farm Safety Review*, January 1957, 12–13.

48. "Radio and Television . . . Award Winners," *Farm Safety Review*, January 1956, 7; "NSC Public Interest Awards," January/February, 1960, 15; "Media Awards Given," *Farm Safety Review*, January 1962, 8; Becky Surface, "Showing and Telling About Farm Safety Week," *Farm Safety Review*, November/December 1974, 9.

49. Gary R. Edgerton, *A Columbia History of American Television* (New York: Columbia University Press, 2007), 12, 77, 89.

50. Ibid. 106.

51. "Farm Safety Field Day," *Farm Safety Review*, July/August 1949, 3.

52. Daniel Kitchen, "Tractor Tipping Demonstrations," *Farm Safety Review*, September/October 1953, 9.

53. Bernard Stodghill, interview by Derek Oden, digital recording, December 29, 2012.

54. *Agricultural Leaders Digest*, October 1953, 18. The popularity of tractor-tipping demonstrations is also evidenced by the number of times Hamilton received requests for information on proper tractor-tipping procedures.

55. Kitchen, "Tractor-Tipping Demonstrations," 9.

56. "New Hampshire Demonstrates Tractor Tipping," *Farm Safety Review*, September/October 1954, 5.

57. The employees of MO-RE, Inc., Bonner Springs, Kansas, composed the advertisement pamphlet entitled "Tractor Safety Teaching Aid." A copy came into Norval Wardle's possession and is retained in his papers; Special Collections, Iowa State University Library, Iowa State University, Ames.

58. "New Tractor Kit," *Farm Safety Review*, May/June 1961, 12.

59. *Transactions: National Safety Congress* (Chicago: National Safety Council, 1953), 36.

60. "National Corn-Picking Contest," *Farm Safety Review*, January/February 1950, 6.

61. Harold J. Schmitz, "Rules That Count," *Farm Safety Review*, November/December 1952, 12.

62. *Transactions: National Safety Congress* (Chicago: National Safety Council, 1953), 37. For a useful discussion of the development and importance of cornhusking competitions in North Dakota, see Gordon L. Iseminger, "North Dakota's Cornhusking Contests, 1939–1941," *Agricultural History* 71 (Winter 1997): 19–45. Iseminger reveals how popular such events were, demonstrating that similar contests would provide an ideal venue for safety activists to communicate their message to a large audience.

63. Charles E. Wilson, "Rural America Cooperates for Safety," *Safety Standards*, November/December 1970, 1. For a detailed discussion of the characteristics of such training programs, the nature of completion certificates, as well as a course curriculum, see Thomas A. Silletto, Dale O. Hull, and J. Clayton Herman, *Safe Operation of Agricultural Equipment* (Ames: Iowa State Cooperative Extension Service, October 1975), 3.

64. Neil O. Snepp, Frank Bobbitt, and Howard Doss, *Vocational Agriculture Training Program* (East Lansing: Michigan State University, Rural Manpower Center, December 1970), ii.

65. Wilson, "Rural America Cooperates for Safety," 2–3.

66. Ibid., 1.

67. Jim Klein, interview by Derek Oden, digital recording, December 15, 2012.

68. L. C. Gibbs, "Pesticide Safety," *Extension Service Review*, January 1969, 6–7.

69. David V. MacCollum, "Tractor Canopies: Army Engineer Research Proves

Their Value for Protection of Operators," *Pacific Builder and Engineer*, October 1958, 3; "Reducing Injuries with ROPS," *Farm Safety Review*, January/February 1973, 14–15.

70. "Safety Is Not an Option," *Des Moines Register* (Iowa), October 27, 1970, 1F.

71. "Zero In on Safety," *Farm Safety Review*, September/October 1970, 4.

72. Page L. Bellinger, *Farm Safety Review*, November/December 1970, 16.

73. "Corn-Picker Programs Pay Off," *Farm Safety Review*, September/October 1953, 5.

74. Ralph E. Patterson, "We Are Responsible for Farm Safety," *Extension Service Review*, November 1965, 14. In 1970, Charles E. Wilson praised the positive effect of the "Cooperative Extension Service and the U.S. Office of Education training programs in safe operation of farm equipment," claiming that of the thirty-five thousand graduates of the program not one had yet been involved in a farm mishap causing "serious injury"; see Wilson, "Rural America Cooperates for Safety," 13.

75. David W. Taber, "Promoting Safety Awareness," *Extension Service Review*, December 1969, 3. Farm safety writers often identified the problem of nonstandardized farm accident research methods, which they believed greatly reduced the usefulness of data collection projects; see "Zero In on Safety," 4.

76. *Transactions: National Safety Congress* (Chicago: National Safety Council, 1962), 8–13; Jack Burke, "Guidelines for Successful Community Safety Programs," *Farm Safety Review*, January/February 1967, 7–10; see also "Are They Getting Your Message?" *Farm Safety Review*, September 1963, 4–7.

77. Robert E. Kowalski, "New Route for Safety Messages," *Extension Service Review*, February 1968, 3.

78. John S. Hitchcock, "The Health of the Farmer," *American Journal of Public Health* 15 (July 1915): 652.

79. James Wilson, interview by Derek Oden, digital recording, December 30, 2012.

80. Bernard Stodghill, interview by Derek Oden, digital recording, December 29, 2012.

81. Bill Tudor, "Outstanding Farmer Works for Future Successes," *Daily Sentinel* (Le Mars, Iowa), March 22, 1966, 3.

82. James Wilson, interview by Derek Oden, digital recording, December 30, 2012.

83. Ibid.

84. Ibid.

CHAPTER 8

1. Kenneth H. Anderson, "4-H Safety Marches On," *Farm Safety Review*, January/February 1949, 3.

2. Maynard Coe, "Youth Works for Safety," *Farm Safety Review,* November 1956, 1, 15; Willie L. Ulich, "Texas 4-H Members Fight Farm Accidents," *Farm Safety Review,* October 1956, 15.

3. Charles and Rose Mary Scott, interview by Derek Oden, digital recording, June 16, 2014.

4. Although most FFA chapters were connected to high school Vo-Ag programs, the organization's national constitution and bylaws indicate there were "collegiate chapters" as well; see National FFA Supply Service, *Official Manual for the Future Farmers of America* (Alexandria, Virginia: National FFA Supply Service, 1972), 5, 23.

5. Ibid., 8, 40–41, 49–51.

6. Ibid., 8.

7. Ibid., 26–27, 63.

8. Ibid., 47.

9. National FFA Supply Service, *Official Manual for the Future Farmers of America* (Alexandria, Virginia: National FFA Supply Service, 1978), 40–43.

10. "4-H Names Safety Winners," *Farm Safety Review,* January/February 1944, 5. Although the development of the 4-H Farm and Home Safety Campaign represented a considerable expansion of 4-H safety activities, the program was not the first. Several state and local 4-H clubs had already autonomously enacted safety programs. For instance, Bob Rupp, an employee of the University of Minnesota's office of publications, indicated that his state's 4-H clubs had been involved in safety activities since 1934. Minnesota youth conducted hazard hunts, distributed safety information, and constructed exhibits prior to the creation of the 4-H national program. For more on these early activities, see Bob Rupp, "Minnesota 4-H Program for Safety," *Farm Safety Review,* March/April 1947, 6–7; and Clara M. Oberg, "They Know Their Hazards: Ten-Year 4-H Safety Program Saves Old and Young Untold Misery," *Farm Safety Review,* September/October 1943, 10–11.

11. "Honor 4-H Safety Champs," *Farm Safety Review,* January/February 1946, 3.

12. "4-H Safety Winners," *Farm Safety Review,* January/February 1953, 3.

13. *Transactions: National Safety Congress* (Chicago: National Safety Council, 1940), 40.

14. Hazel K. Beman, "What's New in Safety," *Farm Safety Review,* January/February 1954, 4–5.

15. *Transactions: National Safety Congress* (Chicago: National Safety Council, 1959), 18. The *Farm Safety Review* featured a significant number of articles whose authors expressed the view that reaching youth could also result in disseminating the safety message to adults. For an example, see "When Youngsters Preach Safety, Adults Listen," *Farm Safety Review,* May 1955, 10–11.

16. Margarite McNally, "Millions of Hazards Removed," *Farm Safety Review,* September/October 1950, 8. Prior to the enactment of this national program, the NSC had encouraged FFA youth to instigate safety awareness activities by

recognizing students who had made special contributions to making their communities safer with certificates of recognition; see "Safety Award for FFA Boys," *Farm Safety Review*, May/June 1943, 12.

17. "4-H Club Safety Winners Named," *Farm Safety Review*, January 1968, 8.

18. John Farrar, "First National FFA Safety Award," *Farm Safety Review*, November/December 1950, 4–5. The FFA national leadership continued to recognize accomplishments and offer prizes throughout the era. By 1956, cash prizes given to youth and chapters totaled $6,100 for that year alone; see "Future Farmers Name Safety Winners," *Farm Safety Review*, December 1956, 1; although national coordination of FFA safety activities emerged in the 1950s, some chapters in states such as Illinois and Wisconsin were already intensively involved in improving safety in their communities; see "FFA Safety Club," *Farm Safety Review*, May/June 1945, 6; and "FFA and Rural School Promote Safety," *Farm Safety Review*, January/February 1949, 8–9, for a discussion of these early state and local programs.

19. Farrar, "First National FFA Safety Award," 4–5; for a discussion of a 4-H chapter in Kansas that was particularly committed to safety, see Franklin M. Reck, *The 4-H Story: A History of 4-H Club Work* (Ames: Iowa State College Press, 1951), 289.

20. National FFA Supply Service, *Official Manual for the Future Farmers of America* (1978), 7.

21. Wessel, *4-H: An American Idea*, 94.

22. Leon M. McNair, "Make Safety Part of the Project," *Farm Safety Review*, March/April 1949, 13.

23. "FFA Corn Harvest Program," *Farm Safety Review*, July/August 1960, 8; "Wisconsin Honors FFA, 4-H Clubs," *Farm Safety Review*, July/August 1960, 10; "Awards for Farm Youth," July/August 1961, 11.

24. "Let the Sun Shine," *Farm Safety Review*, November/December 1969, 13; William F. Steinmetz, "Young Hazard Hunters Wage War on Fire," *Farm Safety Review*, September 1961, 6–7; Joe Van Cleve and Gary Staiger, "4-H Team Up on Bicycle Safety," *Extension Service Review*, August 1966, 20–21.

25. "4-H Club Safety Winners Named," *Farm Safety Review*, January/February 1968, 8–9.

26. "Handbook Out," *Farm Safety Review*, September 1961, 15; "4-H Safety Winners Named," *Farm Safety Review*, January/February 1970, 8–9.

27. Although most of the youth group safety efforts were directed toward more prevalent dangers such as those associated with automobiles, fire, and agricultural equipment, 4-H and FFA efforts included almost the entire breadth of dangers, including those experienced during recreational activities such as swimming and bicycle riding; see Virginia Nance, "Paving the Way to Farm Pond Safety," *Extension Service Review*, July 1959, 149; and "4-H Members Ride Bikes Safely," *Extension Service Review*, March 1951, 42.

28. "Rural Schools Raise Big Crop of Safety Posters," *Farm Safety Review*, July/August 1948, 7.

29. "Safety Displays Win," *Farm Safety Review*, July/August 1950, 5; *Transactions: National Safety Congress* (Chicago: National Safety Council, 1953), 39. For an example of the kinds of advice extension personnel provided in the construction of visual displays, see John H. Behrens, "Open the Way with Visuals," *Extension Service Review*, February 1957, 27.

30. Martha Kohl, "Safety Exhibits Prevent Accidents," *Farm Safety Review*, July/August 1957, 7.

31. E. H. Regnier, "This May Happen to You," *Farm Safety Review*, December 1955, 11; "You Use Your Head: Else You Wind Up Dead," *National 4-H News*, September 1950, 22–23; George Forgey, "Farm of Broken Men," *Farm Safety Review*, July 1955, 13.

32. Regnier, "This May Happen to You," 11–12.

33. "State-Wide Farm Youth Hazard Hunt," *Farm Safety Review*, May/June 1954, 3; Margarite McNally, "Millions of Hazards Removed," 7–8. During the early years of such investigations, 4-H clubs adapted their hazard checklists from those developed by organizations such as the American Red Cross; see "4-H Clubs Use Hazard Check Sheet," *Farm Safety Review*, March/April 1943, 1.

34. *Transactions: National Safety Congress* (Chicago: National Safety Council, 1962), 25.

35. Wardle, "Rural Schools Check Farm Fires," *Farm Safety Review*, July/August 1952, 11; *Transactions: National Safety Congress* (Chicago: National Safety Council, 1952), 16.

36. L. W. Knapp, Jr., "4-H Cubs Learn Fire Control," *Farm Safety Review*, July/August 1958, 12–13.

37. Wardle, "Rural Schools Check Farm Fires," 10–11.

38. McNair, "Make Safety Part of the Project," 12.

39. "Tractor Again Is Farmers' Key Weapon," *National 4-H News*, February 1951, 46.

40. Wessel, *4-H: An American Idea*, 95.

41. William D. Lutes, "4-H Tractor Clubs Build Safe Drivers," *Farm Safety Review*, November/December 1954, 8; Harry Hatcher, "Cutting Corners for Safety," *Farm Safety Review*, June 1955, 4.

42. "4-H Youth Learn Tractor Safety," *Farm Safety Review*, November 1961, 9.

43. "FFA Tractor Program Announced for 1970–71," *Farm Safety Review*, March/April 1970, 9.

44. *Transactions: National Safety Congress* (Chicago: National Safety Council, 1953), 20–21; Art Johnson, "Illinois FFA Tackles Corn Pickers," *Farm Safety Review*, March/April 1953, 6.

45. V. E. Burgener, "FFA Out to Stop Farm Accidents," *Farm Safety Review*,

September/October 1954, 4–5, *Transactions: National Safety Congress* (Chicago: National Safety Council, 1953), 33.

46. Maynard Coe, "An Impressive Example," *Farm Safety Review*, September/October 1958, 1; "Here Are the Winners," *Farm Safety Review*, July/August 1958, 6; James E. Hamilton, "How They Did It in Iowa," *Farm Safety Review*, September/October 1959, 14.

47. "Introducing the FFA Winners," *Farm Safety Review*, January/February 1961, 3.

48. *Transactions: National Safety Congress* (Chicago: National Safety Council, 1963), 8–9.

49. Evelyn Evert, "Demonstrating Electrical Safety," *Farm Safety Review*, November/December 1947, 12–13; Cleva Burks, "'Know How' Wins Trip for Boys," *Farm Safety Review*, January/February 1949, 13. The activities of the Arkansas youth included one FFA student who assisted in improving the electrical wiring for his school; "State Starts Safety Program in Electricity," *National 4-H Club News*, March 1953, 35.

50. "The Miracle of Safety Comes to Montrose County," *National 4-H News*, November 1952, 11.

51. "Progress in Safety," *Farm Safety Review*, November/December 1952, 11. Although the cited farm accident study shows a reduction in farm accident fatalities, it does not provide information regarding the conduct of the study; *Transactions: National Safety Congress* (Chicago: National Safety Council, 1954), 31.

52. "FFA Farm Safety Award," *Farm Safety Review*, November/December 1951, 8; "443 Safer Farms," *Farm Safety Review*, March/April 1951, 3.

53. "National FFA Makes Awards," *Farm Safety Review*, November/December 1954, 7; for additional examples of the results of youth investigation of farmsteads, see "FFA Safety Awards," *Farm Safety Review*, January/February 1953, 4; Peggy Duggan, "Safety Enthusiasts," *National 4-H News*, April 1955, 45; *Transactions: National Safety Congress* (Chicago: National Safety Council, 1953), 25.

54. *Transactions: National Safety Congress* (Chicago: National Safety Council, 1953), 21, 33; Howey is referring to the reality that instances of clogging varied according to crop conditions; increased clogging increased the likelihood of harvesting accidents.

55. *Transactions: National Safety Congress* (Chicago: National Safety Council, 1953), 56.

56. *Transactions: National Safety Congress* (Chicago: National Safety Council, 1955), 36.

57. "Here They Are: The Winners," *Farm Safety Review*, July/August 1958, 6.

58. "A Job Well Done," *Farm Safety Review*, September/October 1959, 14.

59. National FFA Supply Service, *Official Manual for the Future Farmers of America* (1972), 12.

60. See Reck, *The 4-H Story: A History of 4-H Club Work*. Reck reveals that an

enthusiastic group of teachers, administrators, college professors, and government officials were crucial in the development of this national organization.

CHAPTER 9

1. Norval Wardle to Howard Pyle, National Safety Council president, March 6, 1963, Norval Wardle Collection, Special Collections, Iowa State University Library, Iowa State University, Ames.

2. The NSC leadership's response to the idea of the new organization is summarized in "National Institute for Farm Safety," *Farm Safety Review*, November, 1962, 14. Norval Wardle's responses appear in his letter to Howard Pyle, National Safety Council president, May 2, 1963, Norval Wardle Collection, Special Collections, Iowa State University Library, Iowa State University, Ames. Although Wardle indicates that he did not send the letter, it nonetheless reveals his perception of how coolly the NSC received the news of the development of a separate farm safety organization and the corresponding determination of the NIFS specialists to be autonomous. The initial steps to establish a discrete organization had been taken, unbeknownst to NSC leaders, by farm safety specialists at their annual meeting in Florida; see Norval Wardle's group letter, addressed "To All Safety Specialists," September 21, 1962, Norval Wardle Collection, Special Collections, Iowa State University Library, Iowa State University, Ames.

3. Norval Wardle to Howard Pyle, May 8, 1963, Norval Wardle Collection, Special Collections, Iowa State University Library, Iowa State University, Ames.

4. Randall C. Swanson to Glenn I. Prickett, University of Minnesota farm safety specialist, June 21, 1963, Norval Wardle Collection, Special Collections, Iowa State University Library, Iowa State University, Ames.

5. Norval Wardle to Randall C. Swanson, November 15, 1963, Norval Wardle Collection, Special Collections, Iowa State University Library, Iowa State University, Ames.

6. Norval Wardle, "First President's Message," July 15, 1963, Norval Wardle Collection, Special Collections, Iowa State University Library, Iowa State University, Ames.

7. Randall Swanson to Norval Wardle, March 21, 1963, Norval Wardle Collection, Special Collections, Iowa State University Library, Iowa State University, Ames.

8. Norval Wardle to Howard Pyle, March 6, 1963; Randall Swanson to Norval Wardle, March 2, 1963; Randall Swanson to C. E. Stevens, September 19, 1962; all available in the Norval Wardle Collection, Special Collections, Iowa State University Library, Iowa State University, Ames. The NIFS membership application reveals how each of these status levels afforded distinct privileges; Norval Wardle Collection, Special Collections, Iowa State University Library, Iowa State University, Ames.

9. E. W. Foss, farm safety specialist, Cornell University, to Howard Pyle, July 31, 1963, Norval Wardle Collection, Special Collections, Iowa State University Library, Iowa State University, Ames.

10. Randall Swanson to Norval Wardle, October 22, 1962, Norval Wardle Collection, Special Collections, Iowa State University Library, Iowa State University, Ames.

11. Randall Swanson to Norval Wardle, October 14, 1963; and Randall Swanson to Norval Wardle, March 21, 1963; Norval Wardle Collection, Special Collections, Iowa State University Library, Iowa State University, Ames.

12. Don Muhm, "U.S. Could Act if Farm Tractors Not Made Safer," *Des Moines Register* (Iowa), December 6, 1969, 1.

13. Frank Buckingham, "The Dealer's Position in Product Liability," *Implement and Tractor*, August 7, 1974, 61. The view was repeatedly expressed during this era that a shift had occurred in American society placing greater responsibility on manufacturers for design and manufacturing defects and consequent injuries. This view persisted past the era studied, as can be seen in William Johnson, "Product Safety and the Agricultural Engineer," *Agricultural Engineering* 48 (October 1967): 553; T. David McFarland, "On-Farm Accidents . . . How to Make the Farm Machine Man-Environment System Function as It Should," *Agricultural Engineering* 49 (October 1968): 581, 611; and Kenneth F. Packer, Rodney E. Schaeffer, William M. Cade, "Products Liability: How Can We Minimize Hazards?" *Agricultural Engineering* 67 (May 1986): 30–32.

14. Lizabeth Cohen, *A Consumers' Republic: The Politics of Mass Consumption in Postwar America* (New York: Alfred A. Knopf, 2003), 345–346.

15. Buckingham, "The Dealer's Position in Product Liability," 61.

16. Johnson, "Product Safety and the Agricultural Engineer," 553. See "Learning to Cope with OSHA Product Liability," *Implement & Tractor*, December 15, 1972, 9, for additional insights regarding how farm equipment industry leaders were becoming increasingly concerned about the issue.

17. R. O. Diedrichs, "Product Liability: Some Background and a Team Approach to Defense" (paper presented at the 1976 American Society of Agricultural Engineers Winter Meeting, Chicago, December 14–17, 1976), 2. In 1968, E. C. Carlson, an International Harvester engineer, expressed similar sentiments regarding ROPS. He stated, "People are now less inclined to accept safety and safety devices as only 'luxury' items; they now consider safety in the framework of a requirement." See E. C. Carlson, "Tractor Roll-Protection Frames," *National Safety News*, March 1968, 66. In 1973, Warren W. Eginton, a lawyer who specialized in product liability law, characterized the growth of such cases as a "geometric progression in product liability claims"; see Warren W. Eginton, "Minimizing Product Liability Exposure," *Quality Progress* 5 (January 1973): 21.

18. Bill Field, "Getting Personal," *Farm Safety Review*, July/August 1980, 15. Carlton Zink, a veteran agricultural engineer, also provided evidence of these

diverging opinions. In a statement given as part of a panel discussion of the safety responsibilities of agricultural machinery manufacturers at the NIFS 1972 summer meeting, Zink commented, "I am a middle of the roader. Brother Norval gives the manufacturers 'hell'—nicely of course—and Swannie stands up for them and points to what has been done with some approval, but I am not satisfied to take either unqualified stand. I think that there is responsibility enough for all if we are to reach what I consider to be the ultimate goal—the elimination of injuries and fatalities on the farm." Carlton Zink, panel discussion on agricultural machinery manufacturers' safety responsibilities, June 18–22, 1972, 1–2, Norval Wardle Collection, Iowa State University, Special Collections, Ames, Iowa.

19. Speech given by Norval Wardle at a NIFS summer meeting as part of a panel discussion regarding the safety responsibilities of agricultural machinery manufacturers, June 18–22, 1972, 1–2, Norval Wardle Collections, Iowa State University, Special Collections, Ames. Zink, unlike Wardle, had spent most of his professional career in private industry, including a short stint at Firestone, followed by more than eighteen years at John Deere. Randall Swanson was an agricultural engineer and farm safety specialist retiree from the University of Wisconsin.

20. Norval Wardle, "A Safety Engineer's Views on Accident Responsibility," 3 (paper presented at a meeting of the Iowa Trial Lawyers in 1967, specific date not provided), Norval Wardle Collection, Iowa State University, Special Collections, Ames, Iowa.

21. Norval Wardle to Marvin Nichol, February 3, 1978, Norval Wardle Collection, Special Collections, Iowa State University Library, Iowa State University, Ames. Wardle had been working for eight years as a consultant and expert witness, mostly for farm accident victims, since his retirement from Iowa State University.

22. Norval Wardle to Dr. Wesley F. Buchele, May 25, 1978. On August 2, 1977, Wesley F. Buchele had sent a letter to Norval Wardle indicating that he was "attempting to write a history on the agricultural safety movement in the United States," and that he believed Wardle was "eminently qualified to provide information." Norval Wardle Collection, Special Collections, Iowa State University Library, Iowa State University, Ames. See Norval Wardle to Dr. Wesley F. Buchele, August 2, 1977, Norval Wardle Collection, Special Collections, Iowa State University Library, Iowa State University, Ames.

23. Norval Wardle to Dr. Wesley F. Buchele, May 25, 1978, Norval Wardle Collections, Iowa State University, Special Collections, Ames, Iowa.

24. Norval Wardle to NIFS members, May 19, 1986, Norval Wardle Collection, Special Collections, Iowa State University Library, Iowa State University, Ames.

25. Harry M. Philo, "Who's Responsible for Machine Safety?" *National Safety News*, March 1966, 38.

26. McFarland, "On-Farm Accidents . . . How to Make the Farm Machine Man-Environment System Function as It Should," 661.

27. *Transactions: National Safety Congress* (Chicago: National Safety Council, 1963), 27–29. Morris also listed PTO guarding, various features that improved the ease of climbing into the tractor's seat, and more operator-friendly controls.

28. Carlton Zink, "Safety in Farm Equipment: The Manufacturers' Concern," *Agricultural Engineering* 8 (February 1968): 74–75.

29. Farm and Industrial Equipment Institute, *Report of Safety Committee of the FIEI Engineering Group of Committees* (Safety Committee of the FIEI Engineering Group of Committee, 1969), 7, Norval Wardle Collection, Special Collections, Iowa State University Library, Iowa State University, Ames.

30. Wesley F. Buchele, "Product Specifications and Product Liability Litigations," 1, Wesley Buchele Collection, Special Collections, Iowa State University, Ames.

31. Charles Noble, *Liberalism at Work: The Rise and Fall of OSHA* (Philadelphia: Temple University Press, 1986), 46. OSHA's regulatory activities were not only controversial in agriculture, they also attracted critics from a variety of segments of American society. For an example of a particularly passionate critique of OSHA's policies, see Alan Stang, "OSHACRATS: Mr. Businessman, The Buck Stops with You," *American Opinion* (December 1972): 1–13.

32. Glen H. Hetzel, "President's Address" (speech given at the NIFS summer meeting held in St. Louis, Missouri, summer 1985), Norval Wardle Collection, Special Collections, Iowa State University Library, Iowa State University, Ames.

33. For a brief but insightful summary of a public hearing regarding proposed OSHA machine guard standards, see "OSHA in Agriculture: Machinery Guarding Hearing," *Implement and Tractor,* October 7, 1974, 56–58. This piece and others reveal a number of controversies regarding OSHA standards, including the prohibitive cost of retrofitting older equipment with machine guards and other new safety features. For specifics regarding some of OSHA's agricultural standards in the early 1970s, see the *Federal Register* for important documents from the Department of Labor, Occupational Safety and Health Administration, including *Agricultural Tractors: Roll-Over Protective Structures,* 4536–4549 (Washington, D.C, February 4, 1974), and *Guarding of Agricultural Equipment,* 4925–4927 (Washington, D.C., February 8, 1974).

34. Congress, House of Representatives, Investigations and Oversight of the Committee on Agriculture, OSHA Regulations Affecting Agriculture, 94th Cong., 2d Sess. (July 27, 1976), 15–16.

35. Ibid., 2–5. For a discussion of the nature of OSHA regulatory activities in one midwestern state and the controversy the agency engendered among the state's agricultural leaders, see Plambeck, ed., *Iowa Farm Safety in the Twentieth Century,* 143–148.

36. "Farm Employers Must Keep OSHA Records," *Agri-Safety News* 3 (1971): 2.

37. In fact, 88 percent of the respondents either chose "No, a farmer's safety should be his own responsibility and no law is going to make him more safe,"

or "Under no circumstances does the government have any business telling a farmer whether to practice safety" in response to the question, "Should OSHA or OSHA type regulations be extended to include the farmer and members of his family as well"; see "Reporting Back: On Farm Safety and OSHA," *IH Farm Forum* 1 (Fall 1976): 26–29.

38. "Farmers Object Vigorously to Proposed Safety Devices," *Sheboygan Press* (Wisconsin), August 27, 1974, 10.

39. Mary Ovrom, "OSHA Has Some Zingers," *The Hawk Eye* (Burlington, Iowa), August 8, 1976, 21.

40. James Dvorak, "Farmers Not Too Receptive to Increased OSHA Impact," *Lawrence Daily-Journal World* (Lawrence, Kansas), September 29, 1977, 18.

41. Dave Lucy, "May I See You in Your Office?" *Farm & Power Equipment* 54 (March 1972): 40.

42. U.S. Department of Labor, Office of Information, *News: OSHA Names 15 to Agriculture Standards Advisory Committee* (Washington, D.C., June 20, 1972). The committee also carefully consulted such important organizations as the ASAE, FIEI, and the NIFS in determining their recommendations to OSHA officials. OSHA standards also appear to have closely followed the recommendations of these standard writing organizations. See also Rollin D. Schnieder, "The Function and Status of DOL's Agricultural Standards Advisory Committee" (paper presented at the 1973 American Society of Agricultural Engineers Winter Meeting, Chicago, December 11–14, 1973), 1–5.

43. Schnieder, "The Function and Status of DOL's Agricultural Standards Advisory Committee," 5.

44. *Farm Safety Review*, March/April 1968, 14.

45. J. R. Myers and K. A. Snyder, "Roll-Over Protective Structure Use and the Cost of Retrofitting Tractors in the United States, 1993," *Journal of Agricultural Safety and Health* 1 (June, 1995): 185. In 1967, David H. Bucher, a high-ranking engineer for John Deere, indicated that "tractor upsets cause 50 to 70 percent of the fatal farm accidents in this country." David H. Bucher, "A Protective Canopy for the Farm Tractor," *Agricultural Engineering* 48 (September, 1967): 496. For additional discussion regarding tractor overturns as a leading cause of death on the nation's farms, see Ernest C. Carlson, "Frame the Operator for Safety's Sake," *Excavating Contractor* 56 (October 1968): 12.

46. Regarding NSC's Tractor Overturn Prevention and Protection (TOPP), see "NSC's Farm Safety Conference Approves TOPP Program's Safety Objectives," *Farm Safety Review*, July/August 1967, 7; and "NSC Introduces TOPP," *Farm Safety Review*, September/October 1968, 9. TOPP represented the NSC's effort to publicize the benefits of using ROPS and to encourage tractor operators to purchase tractors with ROPS or to refit older tractors with the device. For a discussion regarding the important role that university farm safety specialists had in advertising ROPS's advantages, particularly in Nebraska, see "Demonstrations

Reveal Value of Tractor Roll-Over Protection," *Farm Safety Review*, September/October 1966, 14–15; "Crush Resistant Cab Shows Muscle," *Farm Safety Review*, September/October 1968, 8; and "The Egg That Didn't Break," *Farm Safety Review*, November/December 1969, 4.

47. Sidney M. Wolfe, ed., "Farm Families: An Endangered Species," *Health Letter*, March 10, 1989, 1. Although the lack of overturn protection was a significant problem, it was only one of many safety features missing on machines. In December 1972, Robert E. Glidden, a high-ranking official for Kewanee Machinery & Conveyor, revealed the lack of safety standards in auger design. He stated, "Since there were no written auger safety standards, thirteen manufacturers met in early 1971 with two FIEI representatives and formed a committee. . . . Standards are not easy to write; a majority must agree to them, and some firms might have trouble bringing their products into conformance." See "Learning to Cope with OSHA Product Liability," *Implement & Tractor*, December 15, 1972, 9.

48. Arnold B. Skromme, "The History of Rollover Protection for Farm Tractors" (paper presented at the meeting of the ASAE, Davis, California, January 19, 1988), 4, Norval Wardle Collection, Special Collections, Iowa State University Library, Iowa State University, Ames.

49. "Program Guide for Tractor Safety," *Farm Safety Review*, November 1963, 4.

50. Jack Burke, "Why Wait for Tomorrow?" *Farm Safety Review*, November/December 1969, 10.

51. Robert D. Schnieder and Robert J. Florell, "Farm Tractor Upsets Are Mankillers," *Farm Safety Review*, May/June 1969, 3.

52. "AMA Calls for Tractor Protection," *Farm Safety Review*, September/October 1967, 15.

53. Skromme, "The History of Rollover Protection for Farm Tractors," 2.

54. United States Patent Office, "Top for Tractors," submitted by Henry W. Hanson on November 20, 1922; received patent on August 21, 1923.

55. See United States Patent Office, "Safety Guard for a Tractor Operator," submitted by Osborne Maybrier on July 28, 1952, and received patent on January 3, 1956; United States Patent Office, "Tractor Canopy Guard and Combination," submitted by Earnest E. Selby on February 10, 1956, and received patent on September 10, 1957. In 1953, researchers at the University of California at Davis designed and tested a rollover structure for a John Deere tractor; see Skromme, "The History of Rollover Protection for Farm Tractors," 7.

56. "International Harvester Tests Protective Frame," *Farm Safety Review*, March/April 1967, 14. These manufacturers collectively pooled the results of such studies to develop engineering standards for ROPS. For a brief discussion regarding some of the efforts of John Deere researchers, see David H. Bucher, "A Protective Canopy for the Farm Tractor," *Agricultural Engineering* 48 (September 1967): 496–499, 506. Arnold B. Skromme indicated that as early as 1959 John Deere designers were working on potential ROPS devices. For further

information regarding activities such as those by John Deere, see Skromme, "The History of Rollover Protection for Farm Tractors," 15, 17–19.

57. Bucher, "A Protective Canopy for the Farm Tractor," 506.

58. Carlson, "Frame the Operator for Safety's Sake," 14.

59. Carlton L. Zink, "Anti-Roll Bars: Can They Reduce Tractor Fatalities?" *Agricultural Engineering* 44 (June 1963): 308.

60. E. C. Carlson, "Tractor Roll-Over Protection Frames," *National Safety News*, March 1968, 66.

61. Carlson, "Frame the Operator for Safety's Sake," 14.

62. Carlson, "Tractor Roll-Over Protection Frames," 70.

63. Zink, "Anti-Roll Bars: Can They Reduce Tractor Fatalities?" 308.

64. Myers and Snyder, "Roll-Over Protective Structure Use and the Cost of Retrofitting Tractors in the United States, 1993," 185.

65. Sverker P. E. Persson, "European Experiences with Protective Cabs and Frames," *Agricultural Engineering* 48 (October 1967): 557, 554.

66. Trudy Karlson and Jay Noren, "Farm Tractor Fatalities: The Failure of Voluntary Safety Standards," *American Journal of Public Health* 69 (February 1979): 149.

67. Timothy W. Kelsey and Paul L. Jenkins, "Farm Tractors and Mandatory Roll-Over Protection Retrofits: Potential Costs of the Policy in New York," *American Journal of Public Health* 81, no. 7 (July 1991): 921.

68. Merlin Hansen, "Reducing Tractor Fatalities," *Agricultural Engineering* 47 (September 1966): 472.

69. Carlson, "Frame the Operator for Safety's Sake,"15.

CHAPTER 10

1. Jim Klein, interview by Derek Oden, digital recording, December 15, 2013.

2. John Van Ringelestein, interview by Derek Oden, digital recording, December 28, 2012.

3. Angella Herrin, "Farming Safety Act Proposed," *Daily Dispatch* (Moline, Illinois), December 8, 1988, 1.

4. Hughes, *American Genesis*, 3.

5. Ibid., 5.

6. Herrin, "Farming Safety Act Proposed," 1.

7. Sidney M. Wolfe, "Farm Families: An Endangered Species," *Health Letter* 5 (March 1989): 1. The Public Citizen Health Research Group is a nonprofit organization that researches various public health issues and educates the public about them.

8. Arnold B. Skromme, "A Farm Safety Program Sponsored by Farmers" (paper presented at the annual meeting of the American Society of Agricultural Engineers, Chicago, December 18–21, 1990), 1. Skromme's comments echoed

the point that farm safety leaders had articulated since the movement's very inception: that safety had been given a low priority. He also directed his criticisms at the federal government: "All the agencies of our government, including the lawmakers themselves, have a policy of pushing the farmer harder and harder to produce more and more, by using larger equipment, buying more land, using more fertilizer and chemicals, increasing the vehicle ground speed . . . doing everything to reduce the cost of our food, but at the expense of his health and life! Practically all of the research funded by the federal and state governments covers how the farmer can raise more at a lower cost!" (Ibid., 9.)

9. Burnham, *Accident Prone*, 1.

10. Don Rimathe, interview by Derek Oden, digital recording, January 3, 2013.

11. Pat Larkin, interview by Derek Oden, digital recording, December 29, 2012.

12. Rex Peterson, interview by Derek Oden, digital recording, January 1, 2013.

13. James Wilson, interview by Derek Oden, digital recording, December 30, 2012.

14. Lawrence Schmitz, interview by Derek Oden, digital recording, December 28, 2012.

15. Larry Glenn, interview by Derek Oden, digital recording, December 28, 2012.

16. Wesley Longnecker, interview by Derek Oden, digital recording, January 2, 2013.

17. *Transactions: National Safety Congress* (Chicago: National Safety Council, 1955), 26; *Farm Safety Review*, May/June 1973, 16.

18. Jules B. Billard, "The Revolution in American Agriculture," *National Geographic* 137, no. 2 (February 1970): 169.

EPILOGUE

1. Hurt, *American Agriculture*, rev. ed., 383–388; Crane C. Colt, "Technology Improves Efficiency, Production for Today's Farmers," *Times-News* (Idaho), July 3, 2014, Newspaper Source, EBSCOhost (accessed April 1, 2016), 1–3; Ken Newton, "Maysville Farmer Talks Technology at Congressional Hearing," *St. Joseph News-Press* (Missouri), February 11, 2014, Newspaper Source, EBSCOhost (accessed March 25, 2016), 1–3; Georgina Gustin, "Missouri Lawmakers Pass Bill Backing Large Hog Farms," *St. Louis Post-Dispatch* (Missouri), April 15, 2011, Newspaper Source, EBSCOhost (accessed April 1, 2016), 1–3.

2. Andrea Petersen, "Explainer: Self-Driving Tractors," *Toronto Star* (Canada), August 1, 2015, Newspaper Source, EBSCOhost (accessed April 1, 2016; originally appeared in the *Washington Post*), 1–2; Chris Hubbuch, "Technology, Data Gives Farmers Edge," *La Crosse Tribune* (Wisconsin), September 24, 2014, Newspaper Source, EBSCOhost (accessed March 25, 2016), 2–3.

3. George C. Ford, "Consolidation, Technology Expected to Change Farming,"

Cedar Rapids Gazette (Iowa), March 27, 2010, Newspaper Source, EBSCOhost (accessed April 1, 2016), 1. For a detailed discussion of the historical background as well as some major contemporary aspects of precision agriculture, see K. R. Krishna, *Precision Farming: Soil Fertility and Productivity Aspects* (Waretown, New Jersey: Apple Academic Press, 2013), 1–23.

4. In the first decade of the twenty-first century, the Bureau of Labor Statistics continued to place agriculture on the list of the top ten most accident prone occupations and to place it regularly among the first three most hazardous occupations on the list; see "National Farm Safety Week Has Good Reminders," *Amberdeen American News* (South Dakota), September 20, 2007, Newspaper Source, EBSCOhost (accessed January 19, 2016), 1; and Kelly Stone, "Promoting Farm Safety Awareness," *Agweek* (North Dakota), May 9, 2006, Newspaper Source, EBSCOhost (accessed January 21, 2016), 1.

5. Stone, "Promoting Farm Safety Awareness," 1.

6. Matthew Wilde, "Staying Safe," *Waterloo–Cedar Falls Courier* (Iowa), September 27, 2008, Newspaper Source, EBSCOhost (accessed January 19, 2016), 1.

7. Greg Olson, "Safety Should Always Be Priority, Experts Say," *Jacksonville Journal-Courier* (Illinois), November 8, 2015, Newspaper Source, EBSCOhost (accessed January 14, 2016), 1.

8. Catherine W. Idzerda. "AgrAbility Helps Sharon Farmer Stay in Business," *Janesville Gazette* (Wisconsin), February 21, 2015, Newspaper Source, EBSCOhost (accessed January 19, 2016), 1; Ed Finkel, "The Doctor in the Dell," *Modern Healthcare* 35, no. (March 14, 2005): 2, 4, Academic Search Complete, EBSCOhost (accessed January 21, 2016); Chris Lusvardi, "Agriculture Safety Day Demonstrates to Students the Need to Be Supervised Around Farm Equipment," *Herald and Review* (Illinois), May 16, 2012, Newspaper Source, EBSCOhost (accessed January 14, 2016), 2.

9. Kevin Barlow, "Farm Safety Is Focus of Conference," *Pantagraph* (Illinois), June 23, 2015, Newspaper Source, EBSCOhost (accessed January 14, 2016), 2. The International Society for Agricultural Safety and Health was formerly called the NIFS.

10. "Contributors," Farm Safety For Just Kids, accessed April 7, 2016, http://www.farmsafetyforjustkids.org/thank-you/.

11. Jeffery Meitrodt, "Minnesota Hopes to Replicate Wisconsin's Record on Farm Safety," *Star Tribune* (Minnesota), November 22, 2015, Newspaper Source, EBSCOhost (accessed January 14, 2016), 2.

12. "NIOSH Extramural Research and Training Programs," Centers for Disease Control and Prevention, accessed April 19, 2016, www.cdc/niosh/oep/agctrhom.html.

13. Finkel, "The Doctor in the Dell," 2, 4.

14. Ibid., 1–4. For a discussion of research regarding the potential for hearing damage from farm equipment, see Steven K. Broste, Donald A. Hansen,

Richard L. Strand, and Dean T. Stueland, "Hearing Loss Among High School Farm Students," *American Journal of Public Health* 7, no. 5 (May 1999): 619–622. The focus on cancer was particularly significant since by the end of the twentieth century the National Cancer Institute argued it was likely that farmers' contact with agricultural chemicals made them susceptible to some kinds of cancer; see Hurt, *American Agriculture,* rev. ed., 386.

15. In 1991, the AgrAbility program emerged as a component of the congressional farm bill, consisting of both a National AgrAbility Project (NAP) and the State/Regional AgrAbility Projects (SRAPS), which are closely associated with the land-grant university system. For a more complete description of the history of the national AgrAbility program, see "About AgrAbility," AgrAbility, accessed April 19, 2016, http://www.agrability.org/about/program/.

16. Jeffery Meitrodt, "Tragic Harvest: A Job for Life, and Death," *Star Tribune* (Minnesota), October 13, 2015, Newspaper Source, EBSCOhost (accessed January 14, 2016), 2.

17. Gayle Worland, "Helping the Farmers Who Put Food on the Thanksgiving Table," *Wisconsin State Journal,* November 26, 2015, Newspaper Source, EBSCOhost (accessed January 14, 2016), 1–2.

18. Idzerda, "AgrAbility Helps Sharon Farmer Stay in Business,"1–2.

19. Meitrodt, "Tragic Harvest," 5–6.

20. Olson, "Safety Should Always Be Priority, Experts Say," 1; "MidAmerican Energy Offers 13 Tips to Keep Farmers Safe during Harvest Season," *Newton Daily News* (Iowa), September 24, 2015, Newspaper Source, EBSCOhost (accessed January 14, 2016), 2.

21. Finkel, "Doctor in the Dell," 2.

22. Elizabeth Ahlin, "Tractor Fatalities Trend Up in Autumn." *Omaha World-Herald* (Nebraska), September 30, 2007, Newspaper Source, EBSCOhost (accessed January 19, 2016), 1–2.

23. Meitrodt, "Tragic Harvest," 1–2.

24. Phyllis R. Cotten. "Improving Child Safety Amid the Farm Culture," *Professional Safety* 42, no. 12 (December 1997), EBSCOhost (accessed January 19, 2016), 4.

25. Ahlin, "Tractor Fatalities Trend Up in Autumn," 2–3.

26. Meitrodt, "Tragic Harvest," 2.

27. Ibid. For a succinct and useful discussion of the issue, see Trudy Karlson and Jay Noren, "Farm Tractor Fatalities: The Failure of Voluntary Safety Standards," *American Journal of Public Health* 69 (February 1979): 146–149.

28. Cotten, "Improving Child Safety Amid the Farm Culture," 4.

29. Barbara Marlenga, "Changing the Child Labor Laws for Agriculture: Impact on Injury," *American Journal of Public Health* 97 (February 2007), EBSCOhost (accessed January 21, 2016), 1.

30. Pat Blank, "Proposed Child Labor Rules Could Alter Farm Life," *Talk of the Nation* (NPR), December 28, 2011, EBSCOhost (accessed January 19, 2016), 5; Pam Louwagie, "U.S. Scraps Rules to Limit Jobs Done by Kids," *Star Tribune* (Minnesota), April 28, 2012, Newspaper Source, EBSCOhost (accessed January 14, 2016), 2.

31. Orlan Love, "Farmers Fear Effects of Proposed Child Regulations," *Cedar Rapids Gazette* (Iowa), March 1, 2012, Newspaper Source, EBSCOhost (accessed January 14, 2016), 1–2.

32. Louwagie, "U.S. Scraps Rules to Limit Jobs Done by Kids," 1.

33. Ibid., 2.

34. Ibid., 1.

35. Jermaine Pigee, "Seven Farmers Bring in Harvest for Ailing Friend," *Daily Gazette* (Illinois), November 4, 2014, Newspaper Source, EBSCOhost (accessed January 19, 2016), 1–2.

INDEX

IOWA AND THE MIDWEST EXPERIENCE

The Archaeological Guide to Iowa
BY WILLIAM E. WHITTAKER, LYNN M. ALEX, AND MARY DE LA GARZA

Carnival in the Countryside: The History of the Iowa State Fair
BY CHRIS RASMUSSEN

The Drake Relays: America's Athletic Classic
BY DAVID PETERSON

Dubuque's Forgotten Cemetery: Excavating a Nineteenth-Century Burial Ground in a Twenty-First Century City
BY ROBIN M. LILLIE AND JENNIFER E. MACK

Duffy's Iowa Caucus Cartoons: Watch 'Em Run
BY BRIAN DUFFY

Equal Before the Law: How Iowa Led Americans to Marriage Equality
By Tom Witozky and Marc Hansen

Iowa Past to Present: The People and the Prairie,
Revised Third Edition
BY DOROTHY SCHWIEDER, THOMAS MORAIN, AND LYNN NIELSEN

The Iowa State Fair
BY KURT ULLRICH

The Jefferson Highway: Blazing the Way from Winnipeg to New Orleans
BY LYELL D. HENRY, JR.

The Lost Region: Toward a Revival of Midwestern History
BY JON K. LAUCK

Main Street Public Library: Community Places and Reading Spaces in the Rural Heartland, 1876–1956
BY WAYNE A. WIEGAND

Necessary Courage: Iowa's Underground Railroad in the Struggle against Slavery
BY LOWELL SOIKE

On Behalf of the Family Farm: Iowa Farm Women's Activism since 1945
BY JENNY BARKER DEVINE

The Sacred Cause of Union: Iowa in the Civil War
BY THOMAS R. BAKER

The Small-Town Midwest: Resilience and Hope in the Twenty-First Century
BY JULIANNE COUCH

A Store Almost in Sight: The Economic Transformation of Missouri from the Louisiana Purchase to the Civil War
BY JEFF BREMER

Transcendental Meditation in America: How a New Age Movement Remade a Small Town in Iowa
BY JOSEPH WEBER

What Happens Next? Essays on Matters of Life and Death
BY DOUGLAS BAUER